UPSIDE-DOWN GODS

MEANING SYSTEMS

UPSIDE-DOWN GODS

Gregory Bateson's World of Difference

PETER HARRIES-JONES

Fordham University Press : New York   2016

Fordham University Press has no responsibility for the
persistence or accuracy of URLs for external or third-party
Internet websites referred to in this publication and does
not guarantee that any content on such websites is, or will
remain, accurate or appropriate.

Fordham University Press also publishes its books in a
variety of electronic formats. Some content that appears in
print may not be available in electronic books.

Visit us online at www.fordhampress.com.

Library of Congress Cataloging-in-Publication Data
available online at catalog.loc.gov.

Printed in the United States of America

18 17 16   5 4 3 2 1

First edition

CONTENTS

A BRIEF BIOGRAPHICAL CHRONOLOGY OF GREGORY BATESON

1904	born (May 9) in Grantchester, Cambridgeshire, England; son of William Bateson (b. August 8, 1861) and Caroline Beatrice Bateson (b. 1870?)
1913–17	attends preparatory school at Warden House, Deal, Kent
1917–21	attends public school at Charterhouse, Godalming, Surrey
1922–6	attends Cambridge University: honors degree in natural science and anthropology (1926, under A. C. Haddon)
1926	death of William Bateson (father; February 8)
1927–30	anthropological fieldwork among the Baining and Sulka of New Britain and among the Iatmul of New Guinea, where he first meets Margaret Mead
1930	M.A. in anthropology
1931–37	Research Fellow, St John's College, Cambridge; works on *Naven*, which is published in 1936
1936	marries Margaret Mead (b. December 16, 1901) on March 13
1936–38	anthropological fieldwork with Margaret Mead in Bali
1939	birth of daughter, Mary Catherine Bateson (December 8)
1940–42	works for the Committee for National Morale
1942–43	film analyst, Museum of Modern Art, New York
1943–45	staff planner and regional specialist for Southeast Asia, for the US Office of Strategic Services (overseas in Ceylon, India, Burma, China)
1946–48	attends first Macy conferences on feedback mechanisms and circular causal systems in biological and social sciences; visiting professor of anthropology, New School for Social Research (New York City) and Harvard University

1948–49	research associate with Dr Jurgen Ruesch at the Langley Porter Clinic, University of California Medical School, San Francisco
1949–63	ethnologist, Veterans Administration Hospital, Palo Alto, California
1950	divorces Margaret Mead and marries Elizabeth Sumner
1951	birth of son, John Bateson
1954–59	director, research project on schizophrenic communication under a grant from the Josiah Macy, Jr. Foundation
1957	divorces Elizabeth Sumner
1961	marries Lois Cammack
1963–64	associate director, Communications Institute, St Thomas, Virgin Islands; works with John Lilly on dolphin communication
1965–72	associate director, Oceanic Institute, Waimanalo, Hawaii, and part-time visiting professor, University of Hawaii; begins formal work on ecosystems as an extension of his notion of "ecology of mind"
1969	birth of daughter, Nora Bateson (April 13)
1972	publication of *Steps to an Ecology of Mind*
1972–78	visiting senior lecturer (part-time in the Department of Anthropology), University of California, Santa Cruz
1976	appointed to the Board of Regents of the University of California
1978–80	scholar-in-residence, Esalen Institute, where he works on a manuscript entitled "Where Angels Fear to Tread"
1979	publication of *Mind and Nature: A Necessary Unity*
1980	dies (July 4), Zen Center, San Francisco, California

Adapted from Rodney E. Donaldson, *Gregory Bateson Archive: A Guide/Catalog*, 4 vols. (Ann Arbor, Mich.: University Microfilms International Dissertation Information Service, 1987), 1:5–9.

The pattern is the thing.

What I have tried to do is to turn information theory upside down to make what the engineers call "redundancy" [coding syntax] but I call "pattern" into the primary phenomenon. . . . If a pattern is that which, when it meets another pattern, creates a third—a sexual characteristic exemplified by *moiré* patterns, interference fringes and so on—then it should be possible to talk about patterns in the brain whereby patterns in the sensed world can be recognized.

I believe that much of early Freudian theory was upside down. At that time many thinkers regarded conscious reasoning as normal while the unconscious was regarded as mysterious, needing proof, and needing explanation. Repression was the explanation. . . . Today we think of consciousness as the mysterious, and of the computational methods of the unconscious, e.g. primary process, as continually active, necessary and all embracing. These considerations are especially relevant in any attempt to derive a theory of art or poetry.

Before Lamarck, the organic world, the living world, was believed to be hierarchic in structure, with Mind at the top. The chain, or ladder, went down through the angels, through men, through the apes, down to the infusoria or protozoa, and below to the plants and stones. What Lamarck did was to turn that chain upside down. He observed that animals changed under environmental pressure. He was incorrect, of course, in believing those changes were inherited. . . . When he turned the ladder upside down, what had been the explanation, namely the Mind at the top, now became that which had to be explained.

—GREGORY BATESON

INTRODUCTION

A Search for Pattern

> The Pattern is the thing.
> —GREGORY BATESON

Gregory Bateson has been acclaimed as one of the most influential thinkers of the twentieth century. He was without doubt a transdisciplinary thinker, a necessary prerequisite for any writer putting the case for holism. He moved from discipline to discipline during his life, first as a biologist under the watchful eye of his own father, William Bateson—a famous biologist in his day—next as an anthropologist, then as a social psychologist, and communications theorist—one of the founding members of cybernetics—then as a psychotherapist, and finally as an ecologist. He also helped introduce the concept of feedback as a dynamic aspect of all natural organization. Feedback in Bateson's view provides a means through which it is possible to trace the changing form of pattern in the living world. This is because natural systems are in continual movement; as they oscillate, they are self-organizing and, at the same time, are self-correcting in their dynamics of organization. Feedback refers to their self-correcting dynamics.

His transdisciplinary approach with its strong emphasis on feedback and the nonlinearity of cultural, psychological, and biological order led him to believe that a great deal of previous writing on the question "What is life?" had been poorly posed. All too often the question had become answered in terms of life-as-machine. To Bateson, life was not a machine; there are fundamental differences between a mechanistic and a nonmechanistic understanding of existence, in that life is self-organizing. The question "What is life?" had also been posed in terms of the separation of "internal" and "external" aspects of life, a separation between "withinness" of specific biological characteristics, genes, and "external" laws of randomness and "natural selection." Bateson believed this had created a dualism that leads to an inaccurate account of how evolution creates variation. We required a greater understanding of how interactions between organisms and their environments, between the "within" and the "without" of an ecological setting as it gives rise to evolutionary change. In short, Bateson challenged Darwin's conception of natural selection and took issue with some of the central dogmas of genetics of modern molecular biology as both too positivistic and mechanistic. He wished to initiate a postgenomic vision through a better understanding of pattern and systemic feedback in biology.

The answers to the question "What is life?" had, in the last two centuries, generated too great a reliance on human exceptionalism, and along with this it had placed too great an emphasis on cognitive aspects of the human mind. It had pushed to one side the fact that all organisms have a capacity for perception and that they derive meaning from their perceptions. Furthermore, heavy emphases on the distinguishing aspects of human consciousness, purpose, and meaning turned upside down the primary importance of nonconsciousness inherent in self-organizing properties of living systems. A reevaluation of the psychology of nonconsciousness was needed, and along with it a new understanding of "habit" and "instinct."

Bateson was an original. The noted art critic John Berger states that true originality is never something sought after, or, as it were, signed by the originator as an "original," but is rather a quality belonging to something "touched in the dark and brought back as a tentative question." True originality requires "border hopping," that is, the transgression of existing boundaries of thinking so that "when an insight brought back from an intuitive foray seems to stand up, hold water, or prove its spaces, we are in the face of an original work" (Berger 2005, 87). Bateson crosses at least two of such borders, the first from materialist and/or energy based boundaries of social explanation to that of information and communication as motivation for humans and their wish for relationship. The second border he crosses is an expansion of this approach of relational ordering among human beings to the whole of the living systems. One chapter of this book refers to "the semiosphere" as that sphere of meaning which biologists and other scientists have overlooked while they concentrated on their study of the biosphere. The argument is that there can be no biosphere without semiosphere.

Feedback is a term commonly used today in all aspects of communication, but in the immediate aftermath of World War II, it was both obscure and difficult to understand. It arose out of design for military and technical systems where the electromechanical insertion of a means of adjustment proved to be quicker and more effective than any adjustment made by human beings to antiaircraft guns. The engineering branch of cybernetic feedback helped develop Artificial Intelligence, the study and design of intelligent agents. The field claimed that intelligence could be investigated, and described sufficiently well, that it would be possible to simulate human intelligence by means of a machine. Its version of intelligent agents was of a computational system that perceives its environment and takes actions that maximize its chances of success.

Some of the early initiators of the science of feedback, Bateson among them, also saw that these automatic processes of self-correction involved a whole range of phenomena other than robotic machines embedded in the military-industrial complex. They saw that the principle of feedback, or self-correction, applied in the motor systems of animals and in animal neurophysiology as well, and that feedback, in its broadest reach, could be regarded as a principle of life itself, a hitherto missing process necessary for the survival of living systems.

The details of Bateson's objection to Darwin's ideas of evolution requires detailed discussion, since they are complex, but his opposition to Darwin in no sense supported creationism, the idea of a single creator as a designer of nature, nor were they a rejection of evolution as a concept about change in the earth's biological organization. Instead, he believed that Darwin's notion of natural selection was inaccurate if it is considered to be a universal law of continuity in evolutionary change. Bateson believed that all explanation of living systems requires a nonlinear approach and that any evidence of evolutionary change must include "jumps" or discontinuities (biology now recognizes that "punctuated equilibrium" is a better description of evolutionary change). He also believed that qualitative evidence of a wider set of events, including qualitative evidence of environmental change, must enter into discussion of evolution, for he proposed that the fundamental unit of evolutionary change is not the organism alone but the organism *together with* its environment. In addition, he believed that Darwinian explanations were too reliant on quantitative evidence (statistics) for their justification of Darwinian events.

Bateson also opposed behaviorist assumptions dominant during his lifetime. Behaviorists were always reductionist, attempting to derive a single pattern of events from multiple patterns of events. By contrast, the embodiment of any change, according to Bateson, had different effects at different *levels* of behavior. For example, changing a long-standing habit rarely results in change in a singular behavioral activity but reverberates much more widely.

This is how feedback and feedforward enter into Bateson's approach. As he would explain, changes in any pattern of events affect complex levels of relationship with new patterns of feedback occurring at many different levels, all of which ramify throughout that system of relationships. A very simple example of this would be that of throwing off old "habits" and becoming a "new man" or "new woman." This invokes multiple changes in bodily feeling and changes of perception and a changed view of patterns of interaction, so inducing new perspectives of self-understanding. In its widest reach, change can lead to new patterns of learning. Bateson's approach turned upside down the simplified, clear-cut outcomes that behaviorism sought to obtain. Like other postindustrial writers of his period, his approach to a world of difference sought to liberate thinking by contesting some key methods through which Western science derived its truths, namely reductionism, the use of binary oppositions (dualism), and a nontemporal logic.

Nevertheless, cybernetics alternatives fostered very different practices and perspectives about to these new ideas of information, communication and control. One wing of cybernetics, Artificial Intelligence, promoted an inflated view of its ability to engineer a "mind." From Bateson's point of view, Artificial Intelligence was unable to deal with the mundane world of experience and render it "meaningful." The learning of living systems occurs through experience, and experience yields patterning of events that can be expressed through perception. Moreover, all living systems have perception plus intelligence sufficient to discriminate through use of their perceptions.

In the 1930s, when Bateson began to talk about pattern, nearly all social sciences were encased in suppositions of a mechanical universe, with their ideas of stability and change drawn from Newtonian mechanics. After the war, Bateson believed that if the sciences continued treating the natural world as mechanics, whether Newtonian or quantum, there could no longer be any "science of decency" (Chapter 2). The effects of the detonation of the atomic bomb at the end of World War II may have been a triumph of theory fostered by Einstein, Rutherford, and quantum mechanics, but it brought with it a deep disquiet about the future of applied sciences. Bateson's first spouse, Margaret Mead, shared this disquiet, pointing out in her autobiography that the explosions of Hiroshima and Nagasaki changed everything: "At that point I tore up every page of a book I had nearly finished. Every sentence was out of date" (Mead 1972, 296).

The introduction of nuclear technoscience had made possible levels of destruction that were, by comparison to other eras, infinite. Both Mead and Bateson believed, perhaps naively, that many in the natural sciences would abandon their research in the physical sciences and move into the humanities. Instead, they became witness to the triumphs of physics leading other sciences away from exploring qualitative notions of truth, trust, and beauty, and throwing their intellectual support toward the military-industrial complex instead. Along with these changes came a rise in mechanistic epistemology. As social theories became influenced by the physics of destruction, the social sciences began to push aside many qualitative aspects of form and sensibility to the margins of their interest. Supported to a large extent by governments, the changed epistemology of mainstream science to foster military-industrial applications created unstable feed forward, a vicious circle, which Ulrich Beck recently identified as undermining established safety systems of the state (cf. Beck 1992). Mainstream science seemed unwilling to take into account that the immersion of the sciences into sociotechnology, including nuclear technology, instituted an even greater leap into patterns of risk. The latest example of this vicious circle of risk is global warming, a danger Bateson had identified by the mid-1960s.

With Bateson, we come to a thinker who recognizes that if science continues to see its mission in terms of the replenishment of a sociotechnical order without a broader, deeper, and more holistic vision of comparable replenishment of long-term environmental order, then this narrow vision diminishes our chances of survival. Such a narrowing of vision emerges "out of errors in our habits of thought at deep and partly unconscious levels" (Bateson 2000, 495). As the purposes of our consciousness are implemented by more and more effective machinery, transportation systems, airplanes, weaponry, medicine, pesticides, and so forth, these upset the balances of the body, of society, and of the biological world around us and threaten through their reductionism to bring about a pathological loss of balance. Parts of that reductionism were the game-playing strategies that often underlay the military-industrial technocracy's assessments of risk. Bateson argues that human attempts to game-play nature with little or no attention to feedback will rebound against humanity. One example was the nuclear winter hypothesis of the 1980s, which (after Bateson's death) revealed that the military's version of war games did not consider until that time, that the dust and particles thrown up by a nuclear exchange would create huge, long-term environmental damage of a nuclear exchange between the USSR and the United States. Until that time, the military had done research only on anticipated blast effects (Harries-Jones 1985).

There can be no story of Gregory Bateson without mention of his father, William Bateson, who supplied his son Gregory with some key ideas. Gregory Bateson was born in 1904 in Grantchester, England. William Bateson was a very prominent but controversial biologist, the first professor of genetics to be appointed in Great Britain. Indeed, William Bateson coined the term "genetics" in 1905. William Bateson's rise to prominence began with his introduction to the English-speaking world of the work of Gregor Mendel, the Austrian monk who had undertaken a study of biological reproduction ratios (the XX and XY factors) some fifty years earlier. After translating Mendel into English, William Bateson went on to argue that evolution was not one long story of continuous hybridity, as Darwin had proposed, but that, as the Mendelian scheme of genetic inheritance had shown, evolution exhibited discontinuities in its ratios. Moreover, hybrid offspring are sometimes unable to reproduce by virtue of their sterility. Thus, natural selection was not as universal a principle as Darwin had stated. Though the work of Gregor Mendel was drawn into the Modern Synthesis and became part of the central dogma of biology, this merging of Darwin and Mendel still left William Bateson's outstanding objections unanswered. William Bateson's objections to the sterility of hybrids were perfectly reasonable,[1] but his objections were brushed under the rug by the growing influence of molecular biology. His work became ridiculed, and aggressively so. As the biographers of William Bateson put it, "It is one thing to be buried and forgotten. It is another to be buried and have people come from far afield to stamp on your grave. [William] Bateson's work, rather than being ignored, was actively deprecated both by those who had been his contemporaries and by those who came after" (Cock and Forsdyke 2008, 623).

Gregory was one of three sons, and his father had named him Gregory in honor of Gregor Mendel. His two brothers died young, the eldest being killed in World War I. The youngest Bateson remained the hope of his father, but he was always uncertain that Gregory would meet his expectations of a stellar career in biology. Gregory Bateson began his education by being trained in biology at Cambridge University in England, but when the chance was offered to continue in biology after graduating, he switched fields and began his professional life as anthropologist. Cock and Forsdyke (2008) suggest that one of his reasons for this abrupt change of mind was the interference of his mother in breaking up Gregory's prospective engagement to his girlfriend of the time. Another of Gregory Bateson's worries was that, as a biologist, he would always have to reconcile his work in biology with his family legacy. Moving to another discipline would remove such pressures against his future success in an academic career. He continued to believe in many of his father's ideas, and despite his switch to anthropology, continued to owe his conceptual tools and intellectual habits to his father (Bateson 2000, 73).

As an anthropologist, he undertook three separate fieldwork studies during the 1930s, the main one being a study of the Iatmul of New Guinea. He published a book, *Naven*, a title that referred to the name of a transvestite ritual that the Iatmul performed. His contribution to the discussion of the term "culture" in *Naven* was an unusual one. Instead of looking at culture through his own ethnography and proposing his own definition of it, he looked at the anthropologist as participant-observer in creating his or her own account of culture in these small-scale societies. So, in addition to his empirical ethnography of how the transvestite ritual of *naven* contributed to the organization of Iatmul culture, he developed a sophisticated argument about the ways in which the relationship of observer-informant entered into academic accounts of the concept of culture. In doing so, he showed how the fieldworker can easily succumb to "fallacies of misplaced concreteness" by substituting the observer's viewpoint for the participant's viewpoint of his or her culture (Bateson 1958).

Naven challenged the ways in which fieldworker observations of culture become embedded in an academic construction of "cultural structure." His mentor at this time was A. R. Radcliffe-Brown, whose major influence on anthropology was to introduce structural-functionalism into the discipline. Bateson supported Radcliffe-Brown's ideas about the necessity for a structural or systemic view of the notion of culture, but he also suggested that Radcliffe-Brown's functionalist approach to the notion of structure increased chances that ethnographers would fall into the fallacious substitution of observers' viewpoints for the participant's point of view.

In 1936, he undertook a fourth fieldwork study, together with Margaret Mead, on the people of Bali. Shortly before undertaking fieldwork together, Mead and Bateson married.[2] Bateson was then to become a pioneer in film ethnography, filming trance and dance, while Mead undertook an empirical investigation of childhood development. The two worked together in developing the most sophisticated fieldwork techniques of their time, combining notetaking with film and photography. The impending world war brought Bateson and Mead back to the United States in 1938. Bateson managed a quick visit to England before hostilities began. Mead also managed to take a voyage to England since the United States did not enter the war until 1941.

Bateson tried to enlist when the United States entered the war, but his British citizenship proved a barrier to the being accepted for active service in the American military until 1943. During the early part of the war, Bateson's work was mainly concerned with questions of national morale (among his tasks here was the analysis of propaganda in Nazi films). Yet he managed to publish one important paper in *Readings in Social Psychology* (Newcomb and Hartley 1947), which established him as one of the founding contributors to the discipline of social psychology. The collection also contains the writings of Jean Piaget, Bateson's spouse Margaret Mead, and the celebrated psychologist who had replaced Radcliffe-Brown as Bateson's mentor, F. C. Bartlett.

Bateson joined the morale branch of the Office of Strategic Services (OSS), the forerunner of the CIA, when he was finally accepted for active duty from 1943 to 1945. He engaged in "black propaganda," as it was known, against the Japanese.

FEEDBACK AND INFORMATION THEORY

Bateson's work in the OSS left him with a lot of time for reading and enabled him to gain an understanding of the latest research about information theory and communication, which, he could readily claim, was a necessary part of his task in psychological warfare. Bateson's war years are the subject of Chapter 2. Chapter 3 discusses the immediate postwar era, when Bateson became one of the founding fathers of cybernetics. Once again he was moving across the boundaries of his existing disciplines, from anthropology and social psychology into a foray into communication. He approached the study of information and communication through its formal properties, and found that the formal properties of communication events provide a very different vocabulary than found in the study of material or substance. He found that information has the capacity to trigger material response but is not itself material, so that the difference between the two capacities was critical to a study of both information and communication. Organisms in living systems do not respond to information in the same manner as machines, for organisms' interactions are often unpredictable, and unlike "software" embodied in "hardware," the response of living systems is always one of being minimally "aware." Human beings, as with other animals, utilize perception to derive meaning from pattern, enabling suitable response through being able to grasp of changes in patterning.

Like any conversation, where communication flows through response-to-response, messages between the communicating parties arise from meaningful exchange between participants. The most significant feature here is that in any feedback of messages, meaning is "metaphysical" in the literal sense of the term. It had to be interpreted according to the context of communication interaction and not merely to the vibrations emitted through sound. In the metaphysical exchanges of communication, a first-order response of a *signal* becomes transformed into a meaningful pattern, usually referred to as *sign*. Such interaction was quite different to interaction between the bonds of hydrogen and oxygen. Bateson spent much time showing how the dynamics of pattern formation depend upon recognition of context. The path lay through constant immersion in repetitive forms of messaging. Redundancies in messaging created pattern, which are interpreted as context. By contrast information theory of the time treated messaging in terms of vibrating bits of sound overcoming the presence of noise.

From his fieldwork days, Bateson held that patterns of perception constitute high-level macroscopic vantage points, which enable us to ignore much of that microcosmic world that materialist science investigates. In the macroscopic world of perception, ideas give rise to other ideas, symbolic events give re-minders of other symbolic events, and we create analogies. This macroscopic world of perception is a characteristic of all living systems and has to do with why we see in outlines (Chapter 4). Later, Bateson would define information in its perceptual mode as "the difference that makes a difference" to any set of events. The recognition of difference is primary in perception, but also any meaningful information requires contexts through which it may be under-stood. Although perceptual memory is nowhere near as large as cognitive memory, it can recognize contexts. After the war, Bateson pursued his research career as psychotherapist at the Veterans Administration hospital in Palo Alto, California. His focus on pattern and feedback took him along a fairly radical course, one that, by the mid-1960s, identified him with the counterculture of the times (Chapter 4). The counterculture responded to Bateson's proposals for holistic treatment in psychopathology. His alternative approach avoided the top-down, command-and-control procedures typical of the bureaucracies that seemed to dominate their lives. He derived a new propositional logic about contexts of interaction, using his own "logical typing" for therapeutic purposes rather than relying upon conditions of "truth" in the logic of rationality. The new logic did not follow either strict rules of rhetoric nor mathematical argu-ments (Chapter 5). Later still, Bateson would show that a characteristic of life is its ability to learn through perceptual analogies rather than through the logic of rational procedures.

AN ECOLOGY OF MIND

My first book on Gregory Bateson, *A Recursive Vision: Ecological Understand-ing and Gregory Bateson* (Harries-Jones 1995), drew upon seminal ideas from an extensive archive contained in the McHenry Library of the University of California, Santa Cruz. That volume concentrates on Bateson's writing in the last period of his life, as he explored how primary forms of communication are embodied in ecological order. Missing from the format of that book is Bateson's twenty years as an anthropologist before he moved into cybernetics; a second missing theme is how critiques about the formal—rather than the genetic—characteristics of living organisms have prompted a rigorous debate in biology by developmental biologists against evolutionists. Bateson was a developmen-tal biologist. A third missing theme is how others interested in biocommunica-tion have in one way or another developed Bateson's ideas of holism since his death and given rise to semantic and/or semiotic approaches to ecology. All of these themes seem to justify attention in another manuscript.

Despite the substantial archives on Bateson at the McHenry Library, this latter archive misses detailed information about his early period. Another archival record is lodged in the Margaret Mead Collection of the Library of Congress revealing how Bateson drew from his father's work on pattern variation. It also shows how his research drifts from a functional study of culture to search for cultural configuration, a quest that stems from contact with Margaret Mead and Ruth Benedict. Though Benedict was one of the most influential of the North American figures in cultural anthropology, Bateson considered Benedict's study of cultural configuration as a static (synchronic) appraisal of pattern, while he wanted to embrace *pattern* through the dynamics of its (diachronic) changes. The Library of Congress archive also shows the germination of one of Bateson's most prominent metaphors, that of "balance" in relation to cultural patterning. The metaphor emerges from his encounter with the people of Bali and the way in which they pursue a balance of reciprocities in their society. Balinese society is monistic—unlike the dualist states of the Western world, which separate spiritual order from civil order. Also, their "gods" have a very different place in society from that of the transcendent gods of Western divinity. While the gods of caste engender little of the patterns of dominance-submission usually found between higher and lower castes in Southeast Asian societies, fear of sorcery is part of their everyday life. The transformation of gods to demons and sorcerers and their possible reversal from sorcerers and demons back to benign spirits again are a means through which the people of Bali play with their own fate. He is particularly struck with the way in which bodily balance plays against mental insecurity in Balinese life, and how this cultural configuration maintains its monism compared to the dualism of mind and body in the Western world.

He describes the Balinese as continual tightrope walkers, trying to achieve appropriate balance in all situations. "It is an acrobat's enjoyment both of the thrill and of his own virtuosity in avoiding disaster" (Bateson 2000, 174). He noted that while Balinese regard the gods of caste as "inevitable grooves or tramlines in the structure of the universe" (Bateson 1991, 32), they treat their gods of the village as children rather than patriarchs. The belief in the presence of sorcerers and demons, "upside down" gods and their pervasive presence, is only one manifestation of a number of inversions that occur in Balinese culture.

The years of World War II bring the additional dimensions of information and communication into his postwar writing. Through this additional understanding, Bateson is able to shift his critique to dualism in the Western world and tie this to risky applications of science, specifically nuclear warheads. The end of World War II had brought technocracy and material invention to the center of the Western imagination, along with sociotechnical fragmentation, a dependency on quantitative analysis and a flood of terminology from physics invading the social sciences. The possibilities of retaining balance between qualitative and quantitative methodology and of cultural holism against fragmentation (partism, for want of a better word) seemed slim, but cybernetics and systems science seemed to be one way to mediate such trends.

Part II presents how he took his ideas about cybernetic communication into biology, evolution, and ecology. Bateson develops some radical notions when pursuing his metaphor "ecology of mind." He insists that both "ecology" and "mind" refer to the self-organizing aspects of a joined system, and not—as in all technological manifestations—to parts of systems merged together by external agency. As self-organizing phenomena, mind/ecological relations become both the subject and the object of their relationship. Mind, bears an analogy to a complex ecological system, while the formation of an ecological system—far from being an inert resource for the use of human beings—is "mindful." If the two terms of the metaphor, "mind" and "ecology," are perceived as objects or "substance," its meaning as metaphor is obscure. But from the early days Bateson's discussion was centered on "form" and the perceptual recognition of shape. Shapes that seem to resemble one another are not necessarily empty. Their features have common sets of relations and reveal pattern, structure, and difference among those sets of relations.[3]

"Ecology of mind" becomes a call for science to reframe its understanding of the limits of objectivity, in favor of a participant perspective. The participant components of mind/ecology are recursive, and such recursive processes are not specific to mind/ecology. They relate to evolution in general. When evolutionary processes are considered as both their own cause and effect, evolutionary processes are framed in a complete contradistinction to Darwin's externalist thesis of natural selection (Chapter 6).

In A Recursive Vision, I pointed out that Bateson had borrowed a memorable phrase from the writing of a philosopher of semantics, Alfred Korzybski, in which Korzybski declared that "the map is not the territory." By this he meant that what we perceive are not objects in and of themselves but are maps of their relationships. The mapping process derives from ongoing recursive activity, much of which is unconscious, as in habit, but some of which is consciously learned and so enables difference and the contexts of differences to be compared and distinguished. Patterns of difference and their contexts permit memory, learning, and discipline.

Chapter 7 shows how joint authors T. F. H. Allen and T. Hoekstra discuss how subjectivity underlies objectivity in the observer's formal mapping of ecological territory, with the result that any ecological map will differ in assumptions derived at each level investigated. D. H. McNeil joins Allen and Hoekstra in a system science approach to ecological connectivity from a topological perspective. This is important because mapping of ecosystems deals with intangible circularities or recursions, unlike the very tangible studies of organisms in their environment typical of nineteenth-century natural history. McNeil presents a diagram to show how multiplicity of levels (or heterarchy) is necessary to a study of holism. The topological projection was something that Bateson himself wanted to undertake late in his life but unable to do so.

Chapter 7 also discusses authors who have been directly influenced by Bateson's writing on this path to what Brian Goodwin calls "Post Genomic Biology." The views of development biologists have gained adherence in recent years as a result of the hotly contested "evo-devo debate," in which developmental biology has challenged the hyperimportance that molecular biology attaches to the notion of genes as architects, coding agents, and templates of evolutionary processes. Chapter 7 touches upon morphogenesis, a key study within developmental biology, and melds Goodwin's contribution to that of K. S. Thompson.

Chapter 8 then switches to the perceptions of the observed, of organisms in their environment and how they map relationships. Most biologists until a few years ago considered the "lower end" of life to be robotic in its interactions, and the ramifications of this appraisal of living systems are broad indeed. Biosemantics and biosemiotics offer a reappraisal of this deterministic attitude, for they insist that the study of organism in environment can only be approached through recognizing that the nonhuman animal world, like human beings, acts in terms of its own appraisal of its eco-niche. Ruth Millikan's biosemantics considers how animals use both their communication and perceptual capacities to organize ongoing adaptations. As Bateson himself had argued, they have the ability to take their practical experience of communication, gained through interactions with others of their species, and map patterns of these interactions. The pattern derives from recursive activity in their own locale and knowledge of a variety of relationships in their locale preclude any need for the suppositions of rationality and/or conscious choice that scientists and philosophers demand for "intelligent" animal behavior. Animal communication and animal perception is sufficient for the creation of meaning and appropriate responses. Biosemiotics is even broader in approach to these issues and supports the notion of semiosphere along with a biosphere. The semiosphere expands the idea of the biosphere to include sounds, odors, movements, colors in cells, tissues, organs, and plants as signs of life existing at many, many levels of living systems and makes the case that signs, not molecules or biochemical signals, are the crucial underlying factor in the study of life.

AESTHETICS AND LIVING SYSTEMS

Julian Barnes writes that the biographer's research is akin to a fisherman having a trawling net, which, when it fills, "then the biographer hauls it in, sorts, throws back, stores, fillets and sells. Yet consider what he doesn't catch: there is always far more of that . . . think of everything that got away, that fled with the last deathbed exhalation of the biographee. What chance does the craftiest biographer stand?" (Barnes 1984, 83).

"The catch that got away" in *A Recursive Vision* was a deeper understanding of Bateson's perception of "difference that makes a difference" and how it integrates with aesthetics. As I discuss in that book, Bateson studied aesthetic expression in the poetry and art of William Blake (1757–1827) in his youth. In the past, Blake has been called a mystical poet, yet Bateson came to the conclusion of most of today's literary critics who view Blake's poetic vision as challenging "reason," and its reductionist formulae, which together narrow a vision of the world. Blake raises issues of unity and duality in a totally imaginative way and shows both why and how explanations derived from "instrumental reason" are unable to develop a holistic appraisal of events. According to Blake, the conceptions of Ratio (reason) in the secular mind of the Enlightenment divided mind from body; the inside from the outside; the world from the self; the outside into ever finer degrees of manipulated parts; and, ultimately, the human from the natural. Rationality as the great divider had become its own unbalanced god, an "upside-down god," a fragmented vision inverting the whole. Blake's critique of these aspects of dualism provides a synoptic set of themes that resonated later in Bateson's own critique of technoscience.

Not so evident in his writing was another tradition that Bateson drew upon, that of Johann Wolfgang von Goethe and R. G. Collingwood. Goethe, like Blake, was one of Europe's most prominent literary writers. Goethe was a scientist as well. Goethe was a passionate spokesman for qualitative methodology, a holistic scientist of nature who challenged the sciences of his day to retract falsely drawn boundaries between "mind" (as inside) and ecology (as outside). As a scientist Goethe is best known today for his theory of color, which is still a foundational approach in the teaching of art. As a qualitative methodologist he had also written: "If we want to attain a living understanding of nature, we must become as flexible and mobile as nature itself" (Goethe 1995, 64).

The notion of the flexibility of the observer in relation to the object of investigation is to be found at each step of Bateson's career. Early on, he calls for the investigation of culture to be considered through the lens of change, and argues, as Goethe did, that the fallacies of misplaced concreteness intrude on an understanding of culture, by substituting observer's conceptions for cultural perceptions of those observed by observers (Chapter 1). Chapter 2 shows his support for qualitative method against quantitative method. Chapter 3 introduces the importance of context as a means for understanding wholes, a theme that is also present in Goethe's arguments. Bateson's point was not so much to borrow from Goethe but rather to find a means through which Goethe's approach could be incorporated as part of the epistemology of modern science. Bateson's association with the foremost social psychologist in the United States, Kurt Lewin links back to Goethe's prominent interest in perception and *gestalt*. The latter is the German term for shape, or form, or a configuration. Both Bateson and Mead had been working on this idea of the configuration of cultures in their studies of New Guinea and Bali. Lewin, along with others, tied gestalt together with psychological processes generating action fields (Chapter 3). Bateson became impressed with the way social psychologists in the 1950s stressed the gestalt idea of perception as an active process involving selection, inference, and interpretation, and linked perception to anticipation, or readiness to select particular features of sensory information and ignore others.

Goethe was also a developmental biologist with an interest in morphology. Bateson was a developmental biologist through his association with his father, and he believed that an understanding of morphogenesis was a much more suitable way of investigating part and whole in nature than through its genetic components (Chapter 7). Another Goethe argument was about the means through which nature creates its interrelations of part and whole. As Goethe argues, formative nature expresses itself from all sides and all directions in a sea of connectivity, so that it is no longer accurate to think of connectivity solely in terms of purpose and intention as is so prominent in human experience (Goethe 1995, 55–56).

German scholarship took many of Goethe's scientific themes and incorporated them into the philosophical study of phenomenology, but, as we shall see in Chapters 5 and 6, Bateson stayed with Bertrand Russell's notions of sensibility, modified by cybernetic notions of nonlinear order. Early phenomenology took up perception as a means for aiding the intellect; but Bateson took up perception as a means for opening up scientific discourse, that is, generating greater perceptual holism as an outcome of scientific findings. He agrees with Goethe that the study of nature requires a change in rational sensibility by no longer placing total reliance on physical measurement as an outcome of investigation. Nor does nature study require dialectical logic (Eckermann 1998, 244). The task, as Goethe explains, is to make fuller use of our senses, wakening in us a new order of inner understanding as we bring our external environment into clearer focus. Goethe's arguments ramify in the writing of a twentieth-century author, R. G. Collingwood.

There are no surprises, therefore, as to why Bateson believes that an aesthetic rendering of ecological pattern would aid in an understanding of the total systemic structure of ecology. It would generate new metaphors, new points of view about our relationships with nature and—eventually—induce new behavior (Chapter 9). Bateson's notion of aesthetics is that while it is a response carried along with the physical structure of the properties of light, it cannot be reduced either to a physical dimension or to a psychological faculty. Aesthetics and beauty incorporate imagination and learned experience. It is informational. The experience of beauty always aids in coining metaphors and analogies in which aspects of the whole can be perceived in the parts, and it is always capable of change. The usual arguments derived from materialist science cannot contemplate an ecological vision of this type since materialism proceeds by building unities from a base that is highly fragmented.

It had become evident to him "that metaphor was not just pretty poetry, it was not either good or bad logic, but it was the logic upon which the biological world had been built, the main characteristic and organizing glue of this world of mental process" (Bateson 1991a, 241). Ecological aesthetics was neither animistic nor divine, but an in-between form of perception that enables the generation of a feeling of awe for the unity in living systems. The aesthetics of ecology provide a visual analogue to narrative, and narrative is one of the most important means through which human beings are able to differentiate between map and territory.

It is easy to bypass the significance of Bateson's redefinition of information as "the difference that makes a difference." Bateson always maintained that perceptual processes, unlike cognition, are almost entirely unconscious. Bateson turned toward an ecological aesthetics to express his overall idea that a grasp of totality, where the smaller constituents of ecology are coordinated by the larger ones, is necessary in an understanding of ecology. An ecosystem has an intransitive order in which parts can reconfigure themselves as wholes as they, in one way or another, change their own ongoing configurations of their experience. They cannot do this without mutual coordination with each other. The part-whole integration of ecology Bateson spoke of as "the pattern which connects," and that pattern exhibits heterarchy, or intransitive order. Only here, at the interface of mutually supporting subsystems, can the pattern of differences and the pattern of constancies be perceived (Harries-Jones 1995, 232).

An ecological aesthetics would enable a glimpse at this process of ecosystem formation and change, in which the whole precedes the part. Here "difference" is to be found and the differences that make a difference are formed. Bateson turned to the neurophysiology of the brain only on few occasions to explain events, the most evident being the difference between "up" and "down" in the cortex, or more specifically, the difference between consciousness and nonconsciousness. Despite his sidestep of the neurophysiology of consciousness, recent work on the architecture of the brain has supported the relevance of Bateson's gestalt perspective. The neurophysiologist Iain McGilchrist supports Bateson's position that all experience is an experience of difference (McGilchrist 2009, 97) and that gestalt is part of the human neurology of the brain. The human eye does not have so strongly marked functional distinctions at the level of perception as birds and other animals, rather each cerebral hemisphere attends to the world in a different way in the case of humans. The right hemisphere underwrites breadth and flexibility of attention, where the left hemisphere brings to bear focused attention. This has the related consequence that the right hemisphere sees things as a whole, and in their context; whereas the left hemisphere sees things abstracted from context, and broken into parts, from which it reconstructs a whole (McGilchrist 2009, 27–31).

Bateson claimed to have disciplined his own perception in order to achieve an enlarged gestalt: "I suppose I see beauty and ugliness through my eyes. But the beauty and ugliness, for me, only exist as witnesses to the processes by which they were generated." Even seeing such an ordinary sight of the forming of buds in the axils of plants can yield conclusions about the "massive network of premises, each of which is self-evident and necessarily true within that network of growth and organization." It yields a glimpse at sequencing and ordering, which often are quite different from geometry, physics, and chemistry, but which are necessarily true within a natural network. In an aesthetic response to a natural rose, Bateson wrote, "I am reminded of and enlightened about by my own mental and unconscious processes by the shapes etc., of the animals and plants" ("A Way of Seeing." MC, Book Ms. Box 5). The rose molded in plastic cannot testify to anything except the process of molding in plastic, while the beauty of the natural rose can testify to a very much larger universe of growth and evolution. The beauty of the natural rose must be relevant to any of the injunctive forms of information that govern growth.

Bateson worried that materialist registers of ecological disturbance would either miss signs of ecological change, or, if they derived their perspective from quantitative rather than qualitative methods, would take an inordinate amount of time to establish such a pattern. Unhappily, Bateson is correct. It took until September 2013, approximately thirty years, for the International Panel on Climate Change (IPCC) to come to a definitive conclusion that human beings were a major source of global warming. Yet Bateson's definition of life as organism plus their environment, establishes initial propositions that anticipate that the first signs of ecosystem change would result through some form of change in the mutual reciprocities of the carbon cycle. The "canary factor," or first signs of change, would occur in the informational order entwining mutual reciprocation, he believed, before visible physical degradation of an ecosystem became evident. Chapter 9 presents a recent worrying example of ecological collapse, a disorder occurring among the insects, particularly butterflies and honeybees, and the dangers to humans such a breakdown occasions when these go-betweens of insects and plants and humans disappear.

PART I

THE ENIGMA OF CONTEXT

1

CULTURE

A First Look at Difference

Gregory Bateson's search for pattern owes little to the models of pattern searches made nowadays of "big data," that is, through computer algorithms searching vast databanks of digitized information. He began his search for pattern using ethnographic data of small-scale societies, communities, existing on islands, or along lakes and rivers or in valleys. These were all typical sites that anthropologists headed for, in order to investigate the social behavior of people, their environment, and their culture. The definitions of what exactly constituted "culture" varied in the journals of anthropology, but outside the discipline, the interpretation of culture was usually simplified to that of the contrast between "civilized" (industrial and agrarian peoples) and "primitive" peoples. "Savages" was also a common language term in these years. Even anthropologists widely used the term "primitive people" before World War II, when trying to convince the "civilized" world of the coherent and complex behavior of the cultures that they investigated. "Primitive" was thought a more respectful term than "savages."

As a young graduate student straight out of Cambridge University, twenty-three years old, Bateson had wanted to go to the Sepik River area. This was partly because that area was midway between the coastal settlements of New Guinea, which had already received attention from German ethnologists such as Thurnwald, and the upper reaches of the river unknown to any descriptive accounts by ethnologists. The whole area was a German colonial territory before World War I but was administered by the Australian government after it. A major concern of anthropology in 1920s and 1930s was to obtain information about indigenous culture before it became "overrun" by white intrusion and assimilation. Both administrative control and missionary activity were intermittent in the upper reaches of the Sepik. The mid-Sepik, Bateson reasoned, would be suitable not only for studying an indigenous culture that the German ethnologist, Thurnwald, had not reported on, but would also enable him to observe and report upon black-white culture contact (Letters, LOC, 16/07/1926).

Bateson's initial plans were waylaid when E. W. P. Chinnery, the colonial anthropologist for the New Britain/New Guinea area, dissuaded him from doing research along the Sepik River because of reports of headhunting raids in the area. He suggested that Bateson do his fieldwork among the Baining, a Melanesian people who live as cultivators in the northeast corner of the Gazelle Peninsula, in New Britain, inland from the town of Rabaul. In return for waiting until the Sepik area was more peaceful, Chinnery promised to accompany Bateson up the Sepik. Bateson was delighted with the offer (Letters, LOC, 16/07/1926).

The Baining had a reputation among both missionaries and administrators for having a "primitive culture and mentality," a colonialist comment on the fact that Baining appeared to them to be a society that was entirely apathetic. As a result, both government and missionaries tended to ignore the culture. Bateson arrived in 1927 and chose a particular village, Latramat, in the central Baining area. He pursued his research for ten months, at first speaking pidgin and then taking pains to learn the Baining language, filling many notebooks with descriptive data. Yet, in Bateson's own judgment, the data that he was gathering was next to useless because he was unable to elicit from them basic information about their genealogies, a prerequisite of standard ethnographic report. He found the Baining to be secretive and to resent him as an intruder when he tried to sleep in their houses and involve himself in their ceremonies. They refused to discuss religion with him; nor would they talk about their relations to each other or their taboos. And when they planned some significant ceremonial, they tricked Bateson to go out of the village while they performed their various ceremonial acts. He complained in letters to his mother, Beatrice Bateson, that any ethnographic information had to be dragged out of a Baining a sentence at a time.

Eventually Bateson gave up and moved on to a neighboring people, the Sulka, where, he told his mother, he only had to mention the notion of ceremony and responses about that activity would start coming at such a rate that: "My difficulty is to keep informants quiet while I catch up writing notes" (Lipset 1980, 129). The contrast could not have been more evident, and Bateson found the Sulka's political organization, marriage system, mythology and art—their painting style of geometrical patterns—to be compelling. Yet, his ability to conduct full-scale research among the Sulka was limited as a result of his inability to speak to them in anything other than pidgin English, the lingua franca of black-white working relationships. He had at least managed to speak with the Baining in their indigenous language, although he found himself unable to pursue the finer points of Sulka culture. At this point, he was infected with a bad case of malaria, which forced him to retreat from the Sulka area back to Rabaul.

When he reached the Iatmul a year or so later, he found them to be very different, and found fieldwork much less despairing, than had been the case among the Baining. The problem he chose to depict among the Iatmul was the way in which small-scale societies developed intergroup tensions. In this period of anthropology, the whole issue of conflict and cooperation in small-scale societies was of wide comparative relevance, not only to anthropology but also to administrators and to intellectual audiences beyond the boundaries of the discipline. There was a large discourse on lineage segmentation in kinship studies and how villages formed and reformed. There was also political and religious segmentation, and myths about their congruence and subsequent segmentation of peoples with different languages. There was also clan and moiety segmentation based on differing historical conceptions of where clan or moiety had originated.

Unusually, Bateson began to pay attention not only to the ethnography of village segmentation, or "schismogenesis" as he termed it, but also to the way in which the ethnographer approached his informants in order to gain insights into the culture under study. As his letters reveal, Bateson began questioning anthropology's descriptive approach. In his letters, he berated the discipline for being immersed in a welter of unending ethnographic detail. Anthropology required fieldwork, but fieldwork meant more than collecting a list of habits and rituals. An example of Bateson's first critique occurs while he was in England in 1935, on leave from his fieldwork among the Iatmul. He joined with some of the rising stars in the field of anthropology, Meyer Fortes and Evans-Pritchard (post–World War II professors at Cambridge and Oxford, respectively), to express his discontent with Malinowski and his group of supporters at the University of London. Others were antagonistic toward Malinowski's lack of systematic approach, expressing disdain for Malinowski's theoretical formulations, but held their critical fire because of Malinowski's overwhelming support for extending anthropology within the British university system.

When Bateson did write to Malinowski, he, too, muted the passion of his objections:

As I see it, where you [Malinowski] emphasize the need for complete delineation of *all* the factors relevant to a total cultural situation, I emphasize the need to consider these factors *one* at a time, comparing the action of the same factor in a whole series of separate situations. Where you endeavour to give a three (or four?) dimensional picture of the cultural situation, I try to give a separate two dimensional picture, and don't worry for the moment about the distortions and simplifications which this entails. . . . My method will look to you like a mass of hopeless distinctions and unjustifiable simplifications, while yours looks to me like a hopeless muddle out of which scientific generalisations can never come. (Letters, LOC, 01/11/1935 or 03/11/1935)

Bateson's critical objections concerned the way in which any holistic account of a society could be achieved. In another letter to Malinowski, he expressed admiration for Malinowski's latest book on Melanesian peoples of the Trobriand Islands, *Coral Gardens and Their Magic*, paying respects to the amount of material collected and the way in which its chronological ordering had shown the "fundamental working togetherness of the cultural system" of the Trobrianders. But had he worked with the same material, he told Malinowski, he would have collected together all the cases in which a given formal pattern occurred, and from a set of such collections, built up either a picture of the logic of Trobriand culture or at least a picture of Trobriand thought (Letters, LOC, 05/11/1935).

The letter to Malinowski is a clear indication of the direction in which Bateson would subsequently move. A couple of years later, Bateson wrote that while Malinowski's *Coral Gardens* was an astonishing piece of virtuosity, it had completely sacrificed any central thesis to its method of presentation. Malinowski's chronological ordering of his ethnography had provided a simple method of exposition through which facts served as background for other facts, yet the immersion of culture in a chronology of events prevented anyone else from picking over bits that they might want to use. For Bateson, it was unclear whether this method justified the evidence. An original native document would be preferable in many ways to the sifting of the material through the collector's brain, for the sifting process ends up being nothing but a nuisance (Letters, LOC, 13/04/1938).

Bateson's criticisms were appropriate, in that the term "culture" seemed to refer at that time to anything that the fieldworker had noticed and written up in his or her case study, with the result that the whole discipline found great difficulty in distinguishing culture from behavior. Fieldworkers in the Malinowskian tradition had difficulty in distinguishing whether or not culture was a universal, or whether "culture" existed simply as a case study they had produced. Other difficulties arose in distinguishing "culture" as a cause of events, or "culture" as customary rituals performed throughout life cycles, in which case culture had some biological underlay (Bennett 1998, 17). In effect, anthropology was enmeshed in a descriptive methodology from which it seemed unable to extract itself, and Malinowski had provided little advance into the theoretical problems surrounding the concept of "culture."

THE FALLACIES OF FUNCTIONALISM

For Bateson, culture is merely a bias in the routines of doing things, a set of practices, in the same sort of way that Pierre Bourdieu would later label as "habitus." Bateson argued that it was a fallacy to propose that each cultural trait represents a single behavioral function, or that a class of traits has a systemic function that determines the functions of cultural behavior. It was another fallacy to collect and classify traits of a culture under such headings as economic, religion, geography, kinship, and so on, for this had the result of creating an unintended fit with the Western European way of thinking about their own society. Malinowski and his pupils' designation of cultural habits as "functions" had led to a portrayal of culture as an interrelated set of mechanisms, much like a machine. Thus there was a mechanism for modifying the sexual needs of individuals, and a mechanism for supplying individuals with food, and a mechanism for the enforcement of norms of behavior.

When fieldworkers create categories, these are not real subdivisions of behavior in the cultures anthropologists study, Bateson argued. They are merely *abstractions*, which fieldworkers make for their own convenience in order to divide one set of fieldwork data from another. In his own book, *Naven*, he argues that fieldwork notes labeled "religion" or "economics" are merely labels for various points of view that the observer adopts to make sense of concrete situations, and not phenomena present in culture. It was important that anthropologists avoid Whitehead's "fallacy of misplaced concreteness," a fallacy that occurs when an observer takes an abstraction as a label of convenience, but then becomes convinced that the label has a concrete "reality" (Bateson 1958, 262–263). Marxian historians exhibit another example of the fallacy of misplaced concreteness when they maintain that economic phenomena are primary over cultural-communicative activities (Bateson 2000, 63–64).

Bateson was to discover later, as he was going through drafts of his own writing and preparing *Naven*, for publication, that his own text contained the same fallacy of reification and misplaced concreteness he had accused Malinowski and others of promoting. He imagined that the terms he had coined, such as "ethos," "eidos," and "cultural structure," had some form of concrete reality, and that most of the analogies he had produced in his draft of his book in fact physical analogies, that is, the terms he had used to describe Iatmul made culture a material part of their existence: "I pictured the relations between ethos and cultural structure as being like the relations between a river and its banks—The river molds the banks and the banks guide the river. Similarly the ethos molds the cultural structure and is guided by it." He had deliberately looked for physical analogies that he could use in analyzing his own concepts making what he believed (at first) to be a real contribution to our understanding of how culture works, and he imagined their physical interactions having a material effect. However, "when one is seeking an analogy for the elucidation of material of one sort, it is good to look at the way analogous material has been analysed. But when one is seeking an elucidation of one's own concepts, then one must look for analogies on an equally abstract level" (Bateson 2000, 83). Later, he wrote that the discovery of his own false reification "was difficult to make because of the arrangement of data—the abstraction . . . were of course essential to my understanding of their customs; and it was difficult to separate my abstractions from theirs" (MC, 15/11/1965, 1024–1023).

Instead of rewriting the book, he decided upon a series of Band-Aid solutions: "I had to tune up the definitions, check through to see that each time the technical term appeared. I could substitute the new definition for it, mark the more egregious pieces of nonsense with footnotes warning the reader how not to say things—and so on. But the body of the book was sound enough—all that it needed was castors on its legs" (Bateson 2000, 85). His warnings to the reader not to regard "structure" as a concrete, material aspect of behavioral interaction and instead to think of concepts as points of view adopted either by the ethnographer or by the native informants. His mentor at Cambridge ignored the warnings. When A. R. Radcliffe-Brown received *Naven*, he replied with a long letter in the style of a "welcome to the club" of structural anthropology: "Your capture of schismogenesis [village segmentation] brings you right into the structural school of social anthropology to which I have great pleasure welcoming you" (Letters, LOC, 12/01/1937).

CONSTRUCTIVISM

A fieldwork research break in 1934 was to prove pivotal in his understanding of how to proceed with his Iatmul project. His project was to link what Bateson referred to as ethos, or affective relations of Iatmul people, to what he called eidos, or cultural premises among the Iatmul. Extensive field notes supported his first ethos paper. On the other hand, when he came to writing up his discussion of eidos, or cultural premises among the Iatmul, he junked his field notes. In an early draft of what became Chapter XV of *Naven*, a chapter entitled "The Eidos of Iatmul Culture" he wrote in a footnote of his draft version: "I did not in the field pay any special attention to the point which I now know would have thrown light on this problem, and the material here presented to illustrate Iatmul *eidos* is produced largely from memory unaided by notes" ("The Eidos of Iatmul Culture," n.d. LOC Box 10).

A subsequent footnote states that the whole thrust of his ideas about eidos (which Bateson also identified as the cultural logic of "ideas") came about through reading a book by F. C. Bartlett. Bartlett was a Fellow of the same college as Bateson at Cambridge, a scholarly link that was of much more importance in Bateson's day than now. The specific passages in Bartlett's book that interested Bateson were all about the recall of names and objects in memory, and how this was related to cultural performance. Bartlett's *Remembering* (1932) argues that societies that specialized in rote learning seemed to require memory supports through an acknowledged reference system—chronological references, for example—in order to ensure accuracy in recall. Other societies employed a variety of social relations as prompts to memory retention. In both cases, ability to recall the social circumstances of events was tied together with a chronological account of individual accomplishment in that society. Bartlett argued that rote learning was sufficient to engender memory recall, but that this process of learning was less efficient than learning through the formation or mental construction of "schemata."

Bateson found that the Iatmul indulged very little in rote learning and had no need of chronological prompts. In addition, any consensus that formed around accomplishments in one debate was not necessarily repeated through a consensus on other occasions. The Iatmul had a series of complex associations of merit embedded within totemic discourse, making it very difficult for an observer to understand how knowledge could be learned and retained.

Bartlett's experimental data on memory and recall had given Bateson a glimpse into mental activity in general, and this helped him to understand why there was so great a ramification of totemic imagery in Iatmul society. The cultural processes of the Iatmul required intricate knowledge of totemic names, and, as he had seen in the field, the men engaged in regular debates about the association of individual people with totemic names. He wrote in *Naven*, "the naming system is indeed a theoretical image of the whole culture and in it every formulated aspect of the culture is reflected" (Bateson 1958, 228). In the field, he had found that Iatmul debates about names revealed complex totemic associations, but the outcomes of each debate seemed to alter according to circumstance, indicating intricate knowledge of totemic association, how it could be learned and how it could be retained.

Bateson felt he could trust Bartlett's account, because originally Bartlett had worked on his thesis about memory with respect to his own fieldwork among the Swazi of the borderlands of South Africa, whose cultural memory was tied, at least among the men, to their cattle. The men expressed most of their social relationships through minute particulars relating to the qualities of their cattle. Bateson well understood the importance of what Bartlett was proposing. Bartlett proposed that human learning abilities are related to memory through a process of inclusion and exclusion along successive intervals, that is to say, through a series of "steps" in which "every step follows necessarily upon every other." As Bateson knew, Bartlett's book was totally against the trend in psychology toward behaviorist interpretations of mental performance.

Bartlett's discussion of memory was also an extension of the scholarly tradition of St. John's College, Cambridge, in rebutting both the theory and practices of behaviorist psychology. This had begun during World War I with the notable work of W. H. R. Rivers, who, like Bartlett, was part anthropologist and part psychologist.[1] During the long period in which behaviorist assumptions dominated psychology, Bartlett's book was a minority opinion, but with the demise of behaviorism, sometime around the 1980s, Bartlett's constructivist interpretation of mind became generally acknowledged as one of the most important contributions to psychology during the twentieth century.

Of special interest was the notion of "steps" incipient in Bartlett's work, which Bartlett more fully explicated later.[2] Peculiarly, Bateson never offered a decisive definition of "steps"—perhaps because his references to memory referred to perceptual contexts rather than to cognitive thought, such as Bartlett enunciated. Bateson mentioned the idea of "steps" in his 1936 Epilogue to *Naven* (1958, 272), but it was a very preliminary version. Bateson decided to leave his options open: "The connections of steps in creating *eidos* were not yet 'perfectly clear.'" Was the activation of memory among the Iatmul confined to a few specialists among them? Did it confine itself to particular occasions, especially that of the use of personal names, or did it affect all aspects of *naven* ceremonies, or even the culture as a whole? ("The Eidos of Iatmul Culture," LOC Box 10 n.d.). Nevertheless, he believed that Bartlett had unlocked the "logic" in memory, and the very idea of "steps" as "gaps" in a continuum layered as ordered levels became a central feature of Bateson's later references to "thresholds," "levels," and "boundaries."

SCHISMOGENESIS

Bateson eventually chose to peg his whole ethnographic study of the Iatmul around an extended analysis of forms of relationship that could lead to the breakup of village organization through breaks in affective relations, and their countervailing relations. It was a process he termed "schismogenesis." The novelty of his discussion of schismogenic response is that it is so common in social life. A well-known human experience when an adverse emotional reaction toward another person is left to hang, and not countered in an affective (*ethos*) sense, is that it can lead to more and more adverse, exaggerated, communicative interactions between the respondents.

He identified two forms of potential breakdown, the first of which arose in competitive exchanges in which an initiator of the exchange would attempt to "up the ante" in relations with other members of totemic clans within the village, in a game of unending one-upmanship. He called this *symmetrical schismogenesis*. The second occurred between forms of cultural dominance and cultural submission within the society, and he termed this type of interaction "complementary differentiation in *schismogenesis*." Here, the dominant party to the interaction assumes a somewhat bullying attitude, which leads the other to accept the bullying in such a manner that the dominant one then behaves in an increasingly robust manner, so increasing subservient behavior on the part of the other. Among the Iatmul, typical cases concerned relations between men and women, or between people who stood in a particular kinship relation to each other, or between Iatmul and white administrators. The submissive response was only made in specific circumstances. Members of the submissive group would not necessarily show the same cultural submission toward same members of that group in other activities. Yet differentiation of this sort may become progressive and Bateson found that both types of progressive differentiation could lead to "a progressive unilateral distortion" of relations among the members of both groups, which eventually result in perpetual mutual hostility between them, a hostility that might then precipitate a collapse of relationship and a breakdown of the system (Bateson 2000, 68).

Yet, the Iatmul did not live continually on the edge of a vicious spiral of progressive differentiation, and they seemed to stop short of disorder, for *schismogenesis* contained within it the possibilities for restraint in a volatile situation. Schismogenesis could be avoided through a number of interventions or inverse processes. These inverse processes could be very simple responses. One of these, for example, was awareness that carrying boastful, aggressive behavior too far would result in an understanding by the competing parties that "now that fellow's really done himself in by doing that," or as Bateson puts it "carrying the *ethos* too far."

Other types of processes evoked patterns of consensus building. He explained this complementary behavior as inverting the relationship normally existing between *wau* and *luau* in *naven* rituals, revolving around the polarities of *wau* cultural dominance to *luau* submissiveness. *Naven* was an Iatmul ritual occurring between those males who identified themselves as in-laws to their sisters' sons (*luau*). The ritual was unusual because it evoked behaviors originating from the days not long before, when the Iatmul had been headhunters. If the *lau* achieved a kill, or asserted himself in some way as a result of his bold action, the mother's brother would then act out the statement, "I am a woman" or "mother-in-law" (*wau*) and, putting on women's dress, would then begin a parade through the village. Bateson's interpretation of this act was that it is compensatory, even though the transvestite performance is a vicarious boast. The *wau* is saying, "My *luau* has eclipsed me—I am a fine fellow to have a *luau* like that." The transvestite activity is taken to be a vicarious boast because Iatmul men doubt their power to reproduce children who will become successful headhunters; there is also male competition to become the "mother" of novices at male initiation ceremonies before headhunting activities take place. Schismogenesis therefore was "a process of differentiation in the norms of individual behaviour resulting from cumulative interaction between individuals," that is, a response to a pattern of responses (Bateson 1958a, 175).

Once introduced to cybernetics, Bateson would call the dynamics of ritual compensation performed as *naven* as "feedback." Activity not compensated by *naven* that might quickly escalate conflict piled on conflict leading to village breakup he termed "feed-forward." At the time of writing *Naven*, he admitted that he incompletely understood how the dynamics of inverse processes of feedback dampen down the potential for vicious cycles or runaway, which cybernetics termed feedback, and that each of these processes of response to different forms of interaction that were complementary duals of one another. But he understood the importance of the transvestite dance.

VICIOUS CIRCLES

As noted, Bateson was never confused about "culture," a term his contemporaries associated both with the cultural relativism of their individual fieldwork and with "Culture"—big C culture—as a universal of human life. Culture in the sense of big "C" culture was, in Bateson's view, always associated with patterns. Culture with a small "c" was a bias, a tonus, a rhythm, an accent. A "culture" with a small "c" was that which the anthropologist produced in ethnographic case studies about human social interactional behavior. This might exhibit cultural relativism, as the critics of anthropology always alleged, but could also be understood as natural variance. For Bateson, "Culture" with a big "C" reflected a bias in the universals of human existence, a bias that related not so much to individual behavior, but to the response of individuals in their personal interaction with others. Dominance and submission interactions, for example, were universal. Other elements of universal human behavior were those interactions evolving anxiety, fear, pride, dependence, succoring, and purposiveness. Habit was universal, but as both the Iatmul (and later, Bali) showed, personal interactions may arise from customary or habitual behavior. They were also always dynamic. Their dynamics reveal a second-order patterning, a response-to-response where patterns of response-to-response often differed from first-order events. Both first-order and second-order behavioral patterns were psychological as much as sociological. This led Bateson to the supposition that social anthropology should incorporate psychological dimensions within its field of study, specifically the psychological dimensions of memory.

Bateson stuck with the idea of "Culture" as revealing some universal attributes of human behavior, and he continued to generalize his approach, even when many of his contemporaries decided to stick with comparison of small "c" cultural variations. One example was his postwar conceptualization of schismogenesis, in which he decided to interpret his ethnographic data against that of studies of arms buildup in countries of the Western world. Lewis Fry Richardson, a mathematician who would later provide the first empirical studies of chaos theory, showed that arms buildup and war were much more likely to occur through cumulative interaction between individuals, a second-order patterning, rather than through rational policy. The arms buildup process yielded processes of differentiation leading to vicious circles of interaction (Richardson 1967). Against the usual suppositions about war as a rational policy taken in the interests of a nation, and only undertaken as a last resort, Richardson showed that the causes of war were brought about by such things as psychological attitudes and moods between and among actors in different states. Bateson thought that Richardson's study was reminiscent of out-of-control schismogenesis among the Iatmul, to which the conception of "balance" that he and Mead worked out of their fieldwork in Bali was a counterpoint.

Bateson was quick to engage in correspondence with Richardson over issues of complementary schismogenesis (MC, 12/12/1966, 1068–2a). He learned that patterns of symmetrical schismogenesis are different from patterns of complementary schismogenesis at a second-order level, namely that of response-to-response. In the case of symmetrical schismogenesis, two nations affected only by rivalry, cost and fixed grievances can form a system that is necessarily stable. But in complementary schismogenesis—that of dominance-submission—the origin may be unstable, and further chances of stability can only depend upon creating whirlpool points that circulate in the opposite direction. Here is a case of the general event, arms races, and the (small "c") cultural events of schismogenesis, matching one another. They merge in the more general, but more abstract cybernetic conceptualization of feedforward, as a progressive differentiation in interactions that can lead to vicious circles. Feedback is the counterpoint to feedforward, and appropriate use acts to balance difference in interactions.

One other prominent feature that emerged from Richardson's overall approach was his emphasis on the importance of the types of pattern generated through interactive communication for any pattern of interaction in an arms race constitutes a single unit of communicative interaction, rather than two singular actions separately conceived. The "Richardson Process" also pointed out the importance of perception in the cumulative interaction process. It is the perceived hostility, not objective hostility in any particular message content that is important. Image formation about hostility and the mutuality, the hate or trust, is foremost in these communicative interactions. Finally, Bateson learned from Richardson that it is the forward percept of what might happen that shapes attitude and action in the symmetrical exchanges of arms races, and that in Richardson processes the percept of hate or trust is more efficacious in outcome than rational or self-maximizing thinking.

As we shall discover in Chapter 3, cybernetics would add a whole new layer of understanding about this in its study of circular causal systems. He wrote to Radcliffe-Brown after the war:

> I was remarkably close to the answers [about circular causal systems] with schismogenesis—analyzing this as a circular system in which variable A promotes B, which promotes A etc—in the best Marxian manner. But both I and Marx slipped in failing to see that there also exist circular causal systems which instead of being self-maximating are self-correcting . . . and the whole of social science and psychology is thereby placed in an entirely new framework—and the old ego gets chucked out of the window, decapitated by Occam's Razor. (Letters, LOC, 21/08/1946)

BODY/MIND IN BALI

From the very beginning of his Bali field trip, Bateson was struck by the difference between ways in which ethos could be related to eidos. In Iatmul relations, the men were highly competitive and boastful all the time but could be mediated through Iatmul totemic discourse. By contrast, the whole of Balinese culture seemed to be extraordinarily devoid of competitive emotions. Their stress in social interactions arose from spiritual afflictions. The Iatmul question for Bateson had been about potential for vicious circles of competition, which then would lead to village breakup. In Bali, the question of cumulative interaction was how individuals involved in complementary relations—that is relations of dominance-submission—which maintained their emotional patterns unaffected by the terms of their unequal relationship of caste dominance (Letters, LOC, 14/01/1937). In other words, Balinese society was fundamentally antischismogenic.

He wrote to F. C. Bartlett expressing his perplexity:

> In general Balinese seemed to do things because they are the things to do, without any emphasis upon their own emotions. In day to day personal interaction a child is told what it should do, but not how it should feel. As part of the general lack of intensity of affect, a trait which Freud had emphasized, there was a lack of that other Freudian trait, super-ego formation and guilt. The parents do not get angry with their children and if the child (at any rate very small children) gets angry with the parent, it is met with an aloof brushing aside. Dance performances also revealed an absence of intense emotions between persons, or between those persons playing gods, or otherwise referring to gods in their performance. (Letters, LOC, 01/10/1936)

The dance seemed to demonstrate that there was no culturally expected warmth—either of anger or affectation—and this seemed to be the case in interpersonal relations, as a whole in society. If Bateson was to get any understanding of "affect" or ethos among the Balinese, "the tangle had to be analysed." He had to get beyond observational platitudes about the badness or the beauty of the dance, so he asked Bartlett whether there was any psychological justification for considering lack of linked response as being equal to "lack of affect," and for considering these as being specific to the Balinese, or applicable to a more general phenomenon. He also wondered whether there was any justification in his making a linkage between "this lack of emotional stress to the lack of affect which is mechanically dependent on the balanced carriage [of the body]?"

He began to make contrasts of differences of cultural patterns among the Balinese in terms of the usual comparisons made by the ethnographers of the time, evoking overall perceptions of cultural difference together with semantic differentials. Bali was an unusual example where semantic differences, that is, differences of meaning, occurred through body movement as much as through linguistic expression. Bateson noted that Balinese cultural emphasis on body movement and gesture was in many ways an inversion (reversal) of behavior apparent in Western culture. When Balinese undertook work of any sort in ordinary daily life—working in the fields, women pounding rice, walking about—he noted that the body appeared to be beautifully oriented to the job, with no irrelevant tenseness anywhere and with "an extraordinary ease in shifting from one occupation to another." It was as if every movement were so habitual as to be hardly conscious. In their dancing, on the other hand, there was a high level of tension, often leading individuals to enter trancelike states. He noted, "The contrast between dance and daily life is just the reverse of what we [in the West] strive after. Where we labour in daily life with consciously coordinated movement and enjoy dance as an occasion where we watch or perhaps achieve automatic coordination; the Balinese live automatically and dance consciously" (Letters, LOC, 02/05/1936). Bartlett replied to Bateson's hypothesis that lack of affect—and consequent absence of emotional richness in the culture of Balinese—was "mechanically dependent on the [cultural form of a] balanced carriage" by saying that affect was surely connected with the interlinking of bodily response systems, but that it is not mere linkage or nonlinkage of balanced carriage that should be—of concern. "Affect" cannot be a consequence arising from a single linkage. Linkage has always to be defined in a manner involving different groups of muscles. Thus, any linkage between affect responses invokes the whole linkage of muscle groups that links to other muscle groups, and in fact there may be any amount of complex linkage or response within a single response system.

Bateson also proposed to Bartlett that it might be possible to study classification of differences in the emotional stimuli in terms of the cognitive aspects of personality, and thus, through drawing up a typology of personality, he would be able to make a linkage to cognitive premises, or, in Bateson's language, eidos ("ideas"). If this were a feasible premise on which to proceed, it would fall into the same range of enquiry as that which Mead was undertaking with respect to child "learning." Bartlett replied that, even if such a classification of personality could be described as having to do with "thinking," this was not the type of cognition that interests the psychologist. For the psychologist, any such classification as Bateson had proposed would be considered "cognitive" only in a derivative sense.

Body/mind (eidos) "linkage" became a slippery term for Bateson during the whole of the Bali field trip. At that time, Bateson knew nothing of biofeedback, which was a crucial element in deciphering what he was observing. His thinking at that time seemed to be entirely lineal, as if the linkages he was seeking to make between body movement and cultural ideas were formed in the direct performance of bodily activity. The performance itself might exhibit some form of skilling through practiced response, but in his view this was the result of rote learning. The lessons of Bali as to the circularities between mind/body related to perception and learning, were themes that would enter later. They were worked out, after the publication of *Balinese Character* (Bateson and Mead, 1942).

He was stumped after being told that there was no simple "equal sign" between affect, thought, and bodily posture, as he had seemed to think (Letters, LOC, 13/10/1936). He immediately wrote back, a long defensive letter in which he tried to work his way out of Bartlett's critique. He never sent the letter, which was just as well. A year later, he felt much clearer in his mind as to what he was about, for he believed he had found a key to his problems. He was by now deeply into the questions of linked responses of body and mind at several levels. At one level, there was the linked response of body gestures to pleasure and displeasure, as suggested to him by Bartlett. Then there was linkage in cross-cultural contexts, which would provide "what systems of linked response are related to what systems of cultural emphasis." The idea was an important precursor of Bateson's notion of "context" discussed more fully in Chapters 3 and 4, and, equally, to his understanding of the enormous importance of habituation in relation to both learning and the unconscious (in the sense of the nonconscious) mind.

He observed that Balinese learning about posture is curious: the habit of trusting automatic performance to muscles, rather than to conscious thought. This brought the Balinese "a little closer to animal [aspect of humanity] than they [Balinese] like." He proposed that the characteristic rote learning of Balinese body training leads to specific cultural perceptions that, in turn, are repeated in other areas of Balinese life. The training of the muscles starts at a very early age for almost every baby. Much of Balinese teaching is done by holding the pupil firmly, and then pushing and pulling his hands through the required motions until, in the end, his muscles rather than his mind learn to play their part automatically. This sort of teaching leads quite naturally to habitual automatism, so that they can let their muscles automatically go through their proper forms. Such early childhood learning lends great sensitivity and great awareness to the movements and balance of the body.

If he made a return visit to the Iatmul, he felt he would be able to get records corresponding to his collection of photographs in Bali, and thus obtain "more or less comparable records of linked responses from two contrasting cultures with familiar ethos."

UPSIDE-DOWN GODS

From a fieldwork point of view, the Bateson and Mead approach to anthropology in Bali was typical of the period. It was one of extensive micro-observation of cultural behavior in small villages and among individual families. At the time of writing up, these case studies then became generalized and extended to cover a whole culture. Yet, they were the most experienced couple of their era to be engaged in fieldwork together, and they decided to undertake new and complex methods in data collection. Fieldwork material was collected in triple form: namely in a running record of informant behavior in typed note format; a verbatim record of what was said orally, reported on dictaphone and then transcribed by their secretary; and a strip series of photographs (up to an average of one every fifteen seconds) or ciné taken by Bateson himself in support. Bateson reasoned: "This should give us the material to make much stronger statements about ethos and unformulated culture than anything we have, either of us, dared to do so far" (Letters, LOC, 29/03/1938). He further commented: "Our Balinese technique with ciné and camera etc are so different from anything either of us have done before that we hardly know how significant our evidence is till we have some material collected on the same plan from a different culture." In an evident understatement, Bateson noted that the working out of this material "is going to be very serious proposition." He did not know how he would be able to prepare the Bali material if he were required to undertake teaching in the United Kingdom.

Mead also felt that their several innovations in fieldwork techniques through their triple format extended her own formidable powers of transcribing a running record of behavior:

> I never realized as vividly how very dependent I am in my thinking upon having good comparative material always present in my mind. And I can't compare 40 observations on a Manus baby, all merely recorded in words, with 400 observations on a Balinese baby, a good part of which are photographic and verbal records. The levels are so very different. Where before I occasionally made a sample of behavior over time which would run to two typewritten pages for an hour, we now have records of 15 typewritten pages and 200 feet of Ciné and a couple of hundred Leica stills for the same period. The recording is so much finer that I feel as if I was working at different levels from any work that I've done previously. And every time I tried to think comparatively about Balinese materials I was stuck. (Mead 1977, 213)[3]

Their meticulous methodology was an attempt to produce a more rigorous scientific approach through a total commitment to empiricism. Jane Belo, who joined them in 1936, noted that their methods of using photographic stills corroborated with ciné film and meticulous note taking were all synchronized by a chronometer timing of their watches, which were set at the moment they began their observations (Belo 1970, xxv). Yet, the material collected was evidently so massive that it would have required concerted work over the best part of several years to present to scholarly or public audiences, even in normal circumstances. And war intervened. The material never did get published in the form that it had been collected. Initial presentations of their study of Bali were hurried affairs, and more like a visual collage. This was the case with their joint book entitled *Balinese Character: a Photographic Analysis*, which appeared in 1942 (Bateson and Mead 1942). Bateson's self-assessment of his joint book on Bali with Margaret Mead was that their book was "hurried" as well as "descriptive and illustrative. . . . Many of our new hypotheses remain implicit."

Bateson's use of ciné film was oriented to Balinese dance and ritual performance, while his camera gave visuals of social interaction among family members, along with ethnographic portraiture of village life. Bateson praised the dance performers' evident skills, though he pointed out that "the Balinese 'parrot' all their behavior, they do not seem to learn through formal instruction and while undergoing tutoring individuals make little or no attempt to understand what he is learning and scarcely cares for later reward or punishment. . . . They do not have their eye on some distant goal and work towards it with understanding. Rather they live in the immediate moment" (Bateson 1943, "Commentary on Bali Exhibit"). The impulse for learning about dance movement among the Balinese appears to reside not so much in the whole bodily activity, as in each separate limb: "It is as if each separate limb or each joint had a head of its own." The same "unconscious notion" is expressed in many Balinese postures, especially those postures that emphasize the sensitivity and independence of the hand and the fingers. Bateson suggested that this type of learning, with its fragmented but independent parts, fits naturally with motif in Balinese folktales, in which there is often a whole series of supernatural beings who personify parts of the body—heads, arms legs, trunks, etc. Supernatural beings are associated with witchcraft and inhabit spiritually dangerous places such as cemeteries and deep forested areas.

One of Bateson's fascinations with Balinese culture was with the evident contrast between the Balinese response to violence compared to the boastful, aggressive stance of the Iatmul of New Guinea. He presented a summary of these contrasts in an exhibit organized by the American Museum of Modern Art in 1943. Bateson had expressed the hope that the exhibit would combine film, photographic, and observational method shown together, as a multimedia presentation, but Bateson's chance to curate an exhibit of his and Mead's ethnographic work on Bali in this manner were dashed when preparations for the exhibit proved that this was not technically possible. As a result, the exhibit presented only 169 of the several thousand photos that Bateson had taken, with the photos presented in traditional museum style together with Bateson's brief commentary on them.

He and Mead had also collected a sizeable number of carvings and paintings, but only a small number were included in the exhibit. A majority of the

Balinese carvings in the exhibit emphasized problems of balance and movement. He also presented the extraordinary facility of caricature in Balinese
theatrical productions through *wajang koelit* or shadow play puppets. His overall collection of 900 included wooden figures of recent manufacture, together
with paintings of ritual and spiritual events, some of which represented artists
dreams of supernatural figures and some of ritual events, or mythological dra-
mas inhabited by supernatural beings.

The exhibit emphasized the many possibilities to avoid undertaking violent
action. Substitute forms of aggression included kite fighting, cricket fighting,
and, above all, cockfighting: "the man himself stays out of the conflict while
he sends in his [fighting] cock or his [fighting] cricket as a sort of extension of
his own personality or some substitute for himself." Yet, at cockfights, "most
of the time is spent in titillating the cock's anger." His thematic was showing
how all these examples fit in with an overall Balinese distaste for emotional
involvement with others. Instead of engaging angrily with others, individual
men were more likely to turn their aggression inward upon themselves, and in
the case of going readily into trance, "trance is easier to achieve than anger," as
he wrote in his commentary accompanying his exhibit (Bateson, 1943, "Commentary on Bali Exhibit").

Tantrum scenes between parent and child provided another major clue to
the lack of aggression in Bali, where the parent typically instigates a tantrum
which is then either ignored or laughed at. Tantrum scenes were part of a
complex of learning in which "the child learns unconsciously never to respond
when somebody tries to play upon his emotions . . . and learns that anger gets
him nowhere." This particular segment of Bateson's commentary was accompanied by a series of photographs that show "the son's growing misery and
his impotence while his mother gazes off into space and flirts a little with the
other baby. Finally the mother casually picks him up and does not respond
even to his angry pushing at her face."

Bateson linked this commentary to the general observation that the childhood training in not showing emotion is related to the fact that "there is no
oratory in Bali and no [rival] factions." This point that might have been correct
in the historical circumstances of colonial rule in which Bateson had undertaken fieldwork but was certainly untrue in relation to the postcolonial history
of Bali.[4]

The exhibit also portrayed *kris* dancing and linked it to the ceremonial
drama of witch, dragon, and king (Barong and Rangda), performed then many
times a year in its many varieties. Bateson's textual references of the *kris* dance
also linked the inverted violence of the Barong/Rangda dance and to lack of
forms of violence, or lack thereof, in Balinese society as a whole. The plot in

kris dancing depicts a quarrel between the witch and the king. As this conflict develops, the king changes form into a dragon (*barong*) and is aided by young men who suddenly appear, each armed with a *kris*, and attack the witch. The witch then changes form and becomes an enormous replica of Durga, the goddess of Death. The witch waves and laughs at them, and they fall to the ground in an unconscious state. Though the Barong/Rangda begins as a purely theatrical performance, the conflictual parts could often appear to get out of hand. Later, the young men get up "in some middle state of somnambulism" and in this state they turn the daggers against themselves. The aggression that could not be expressed outwardly is turned inwards. The scenes his exhibit depicted, "is one of extreme violence and leads to a climax of mixed agony and ecstasy, after which the men are taken out of their trance state with incense, smoke and holy water."

The attack, Bateson comments elsewhere, can also be interpreted as an incident in which a bride or a mother-in-law is a witch and brings plague and pestilence on the world (Letters, LOC, 12/01/1937). The king who changes his form and becomes Barong may be interpreted as a bridegroom. Barong appears as "a jolly lolloping sort of dragon, with one man in front for the front legs and another behind for the back legs, like a stage horse, but Barong is also a slang term for penis." Today, Balinese culture still gives a central place to *kris* dancing, with abbreviated performances that are available for tourists. They are quite eye-catching, even though the tourist performance has little of the intensity Bateson describes. However, Bateson's sexual interpretation of the Barong/Rangda encounter is questionable for reasons to be explained later.

One of the artifacts presented in the Bali Exhibit was of that of an upside-down god. Bateson's program notes for that occasion stated that making carvings of upside-down demons, of "a low demon" abasing his head, was "an inverted joke on the caste system." The upside-down god was a hobgoblin, a "trickster," as it were, challenging all the gradations within caste organization and the sacredness, or security, of such gradations within caste hierarchy. His "Commentary" in 1943 stated that Balinese people had an obsession with the notion of uprightness and of balance in which the head is always held high, with these postures always matching their observance of caste ritual. It was not enough for the Balinese to memorize that "the head is the sacred part of the body," rather, this is a feeling that must be demonstrated in many different contexts, along with the "joke" of its inversions. The upright head was the center of integration and its concomitant idea of balance, while, by contrast, Balinese linked witches and demons with fragmentation of body parts. Bateson presented other photos of masked actors and small children illustrating this theme. The witch is fear personified; she is both frightening and frightened, and thus echoes the mother's characteristic fear about fragmentation of body parts and abasement of posture.

He wrote later of other relationships in Bali that match the inversions typified by upside-down gods:

> For the Balinese think of relationships between gods and people as between children and parents, and in this relationship it is the gods who are the children and the people who are the parents. The Balinese do not expect their gods to be responsible. They do not feel cheated when their gods are capricious. Indeed, they enjoy minor caprice and charm as these are exhibited by gods temporarily incarnate in shamans. (Bateson 1987, 26)[5]

Other commentators of the time had noted that demons (*léyaks*) are associated with abasement of the head and could be seen in appropriate situations only by bending down and peering through the legs. Miguel Covarrubias called *léyak* a person who through knowledge of black magic can assume absurd, weird shapes to harm others (Covarrubias 1989, 410). Walter Spies reported that a person could be recognized as a *léyak* through such tiny indications as a groove in the upper-lip, an upside-down image in the eye, and/or the failure to look the person you are talking to in the face. At nightfall, the *léyak* haunts the graveyard in the shape of an animal—monkey, dog, pig, wild boar, horse, or tiger—and makes its black power felt. A normal healthy person is not in much danger from *léyaks,* wrote Spies, but when a person's physical resistance is weakened, *léyaks* have the power to destroy them. Both village and temple are protected from *léyaks* by *balian*, who are also wizards, but of a different kind. They are able to go into a telepathic trance during which *balian* can see the cause of illness and its cure (Spies and de Zoete 2002, 88).

The interleaving themes of *léyak* and *balian*, of the ritual drama of Barong and Rangda, of the manifestation of gods as demons, and demons as gods, and *balian* who may be *léyak*, has always been problematic for anthropological interpreters as these forms of transfiguration and counterfiguration are so different to western ways of thinking. The easy route is to categorize these beliefs as forms of witchcraft and magic, and to discuss them in relation to other known forms of witchcraft. Both Bateson and Mead were prone to doing so. Bateson's comments in his Bali exhibit included insights on the emotions and motives that constitute the intricate rules of social life and how these "make the Balinese obey" their intricate rules. Among these, Bateson lists simple fear "to stand—unsupported in the open," meaning the fear to take personal initiatives, as well as the fear of making mistakes, and its ultimate sanction, loss of membership of community.

To these were added the particular cultural attributes of Bali: "the fear of uncleanliness," as every Balinese feels a horror at the animal nature of humanity, and "the fear of incompleteness." The fear of incompleteness is associated with a very demanding notion of "balance." The notion of balance gave a very precise delineation of what constitutes a "complete" family, for example, balance among the sexes of those born to a family. It included completeness in bodily parts, namely the requirement for all babies to have all the fingers and all the toes. The necessity for balance in every aspect of life related, in turn, to the people of Bali having extraordinary muscular awareness in their balance of movement. There were also existential fears attached to the notion of balance: "the fear of disorder." The Balinese child learns that the world "has a safe and reassuring core away from which it is dangerous to wander." And there were the fears of the cosmic order as exhibited in the Bali drama—rituals of Rangda/Barong, witches, and other supernaturals that might bring pestilence and misfortune.

BALANCE

While they were engaged with their fieldwork, Bateson and Mead came to the conclusion that Balinese notions of "balance" could best be expressed through highlighting the phases of *kris* dancing, the dance in which the traditional daggers of Balinese males are drawn in order to inflict wounds upon the self, rather than in fights against an opponent. Yet here joint authorship generated theoretical differences in their interpretation. For Bateson, the phases of *kris* dancing were to be judged in terms of the absence of the process of schismogenesis that he had depicted among Iatmul. Balinese society, unlike Iatmul society, did not engage in this type of rivalry and competition, and nor did their interactions resemble competitive situations in the Western world (Bateson 2000, 107–127).

> And it's not for lack of potentially schismogenic situations. There is plenty of caste, hierarchy, etc.—but no detectable schismogenic strains. The only exception to this is, I think, the relationship between mother and small baby—she titillates it—nice little thing a baby, it wiggles when you poke it and its penis gets up when you wiggle it. So it gets its response mechanically played with from the start and apparently develops a defensive non-reactive posture. Apparently on this base is built the whole non-schismogenic system of the culture—It will be a nice contribution if we can get it phrased properly. (Letters, LOC, 17 /05/1939)

Bateson expressed a hope that the *kris* dancing would become the central focus around which other ethnography was oriented in much the same way that the *naven* ritual had become the focus of his work among the Iatmul. *Kris* dancing seemed to give explicit evidence of how emotional engagement in Bali seemed to tail off or plateau—before any climax occurred. Bateson wrote to Bartlett: "From the very first time that we saw *kris* dancing we knew that the trance was closely linked up with orgasm" (Letters, LOC 12/01/1937), but their initial investigation had left open the question of whether the orgasm achieved was primarily sexual or an emotional plateau. Subsequent investigation seemed to show why the *kris* dance postures seemed to be related to sexual postures. The motions of the two sexes in the *kris* dance were different in their emphasis. His ciné camera had captured on film several extensions of this observation, so there was justification for isolating and classifying a range in postures of other sorts. With the *kris* dance as fundamental to the pattern of posture, he wrote, it would be possible to develop an analysis of the standardized systems of posture-gesture relating both to bodily movement and to cultural training for sexual encounters in adulthood.

The idea of this analysis is that in the *kris* dance, as in other forms of cumulative interaction such as quarrels or activity that might introduce warfare, cumulative personal interaction becomes muted and instead a "continuing plateau of intensity" is substituted for climax, and in childhood "similar sequences have the effect of diminishing the tendencies toward competitive and rivalrous behavior" (Bateson 2000, 113). Though Bateson said he had no clear documentation about climax and plateaus in sexual behavior, the explicit sexual references opened up their fieldwork to Freudian notions of intensity as driving psychological outcomes of interpretations of social bonding. Always a critic of Freud, Bateson had specifically rejected the use of Freud's frustration-aggression hypothesis in the Bali situation. Freudian notions of "frustration—aggression" made absolutely no sense in Bali because Balinese society itself was "almost completely non-schismogenic." Mead, on the other hand, continued to combine her version of role theory with childhood learning of roles arising from the psychodynamics of mother-child bonding, and was much more prone to citing Freudian psychodynamics. Their joint publication, *Balinese Character* (1942), has many suggestive Freudian references.[6]

Bali had provided several turning points in Bateson's intellectual conceptualization of cultural pattern in big "C" terms, and its processes of change. When Bateson had first began to write about change in New Guinea, on the prospect of changing relations between indigenous peoples and their administrators (black and white), he followed the lines of argument used in physics and mechanics about balance and change. Given conditions of physical stasis, a momentum pushing in one direction could be countered by a reversal in the momentum of the push. This image of stability and order achieved by energy transforms when transferred to social organization provides a template for the dynamic of change whereby change moves social conditions from one state to another state. If such a process overshoots desirable stability, reversal of the processes of momentum will achieve balance (Bateson 2000, 66–67). This idea of equilibrial balance, that reverse energy pushes could balance out instability, became a common theme of the study of social change in social anthropology from the 1930s to the 1960s (Wilson and Hunter 1945).

The Iatmul study had shown that there was a very slim border between maintaining stability in society, and running amok. The Bali case showed that balance can be maintained in a highly fluctuating environment that has little to do with mechanical equilibrium and much to do both with value-sense of bodily equilibrium and with response to realm of both play and fear about beliefs in the rotation of gods to demons, and back to gods again.

Returning to themes he regarded as central to Balinese life some thirty years later, he wrote that the relation between body and mind was a special feature of Balinese culture, and so too was the premise that the "gods" are filial rather than parental in relation to mortals. A third important theme overarching theme was the Balinese conception of "balance" in a highly fluctuating and violent spiritual universe, where the values of uprightness, integrity, and holism clashed in social life with fear of meeting sorcerers and upside-down gods. The meticulous way Balinese attended to the cultural premises of "balance" was associated in its cultural imagination with the obverse, the morbid dreams of fragmentation of body parts that accompanied an imagined encounter with a léyak or other sorcerers; so too was their emotional style of nonattachment. His comments were part of a proposal for a film contrasting the premises of cultures in different societies, and he expressed the hope such a film would show how the value system in Bali was "totally different from the west," and would render "visually concrete" these inverse and obverse, but abstract cultural premises, which he and Mead had written about (MC, n.d. c. 1969, 488–3). The film was never produced.

2

A SCIENCE OF DECENCY

Preparing publication of the Bali material coincided with a frantic set of activities as war threatened to break out in Europe. Gregory Bateson and Margaret Mead believed that the American public had little understanding of the threat to democratic institutions stemming from the rise of Nazism or that the fascist threat to democracy would come to the United States sooner rather than later. Bateson was already well aware of the existence of the German eugenics program, its evident political extension, and the path of violence to which it might lead. As a graduate scientist, he also had some understanding of what had gone on in Rutherford's physics laboratories at Cambridge University, and the probable development of nuclear fission from uranium-235. He began circulating his point of view about the dangers of fascism to his immediate peers in the United States, suggesting to them ways and means in which, through their help, the public could become more aware of the threat.

His letters show how much he was committed at this time to the importance of social science, and to his belief that the social sciences would have a great part to play not only in the victory against fascism, but also in waging the peace. Both he and Margaret Mead had discussed the possibility of a general behavioral science emerging from the separate disciplines of anthropology, sociology and psychology. They also recognized that Germany had systematically organized its social science resources in order to conduct propaganda, and that the only response was for both Britain and the United States to undertake the same sort of countermeasures. His first suggestion was for ways and means to analyze shortwave broadcasts from Germany to the United States. During the "phony war" period, in September 1939, Bateson made a brief trip to England to attend to family affairs. While visiting Bartlett and his confreres at the university, he broached similar ideas to the Master of St. John's College

Cambridge, in the hope that the Master would pass his viewpoints on to those in command. As Bateson wrote on the subject: "Every country knows best how to propagandize its own people. . . . If we are to get any sort of hearing in a hostile country we must first soak ourselves in their propaganda. We must be able not only to write idiomatic German but to feel and talk Nazi ideology, to feel German sentiments" (Letters, LOC, 20/09/1939). The hope of getting some sort of hearing was a reasonable supposition, because those in senior positions at Oxford and Cambridge were usually within the corridors of influence. Yet, the whole trip cooled his hopes for any government recognition of the need for professional social science recruitment in Britain in the war effort.

At about the time he made this voyage to his family, the British had begun to set up propaganda and counterpropaganda measures by instituting a Special Operations Executive, the SOE, under Sefton Delmer. His judicious enquiries as to whether he might put his name forward were sufficient to find out that the SOE would not choose to draw its staff officers from professional psychologists or social scientists. The SOE appointments would, in fact, draw from businessmen, writers, journalists, lawyers, and junior politicians. Academics on its staff drew from those associated with upper-middle-class families, or the sons of the elite. Administrators at SOE believed that "where man management, total trustworthiness, and a high intelligence were essential prerequisites," a specially selected elite had to be recruited (Dear 1996, 21). Bateson recognized the trend and came back to the United States disgusted with what he regarded as the lack of professionalism in the British approach. Even so, the SOE was able to engineer several successful propaganda operations early in the war (Delmer 1962).

IMMINENT DANGER

The situation was different in the United States. Here, Bateson had an influential peer group, the social psychologists Kurt Lewin, Abraham Maslow, Larry Frank, and Alex Bavelas, among others, none of whom needed convincing of the necessity for recruiting social scientists if the United States came into the war. To these, one might add Margaret Mead's peer group as well.[1] Erich Fromm and Kurt Lewin both had firsthand experience of the rise of fascism, and Fromm was already beginning to publish his influential research on "the authoritarian personality." Unlike these professionals, the public at large was split on the issue of the war in Europe, with the percentage of the American populace supporting intervention in the war being less than those who maintained a position of isolation and supported staying out. The isolationists' campaign was largely financed by big business, and with opinion running so close the interventionists called their lobbying against the isolationists "the battle for the USA."

Among those proposing intervention were small groups of publicists and radio commentators who had some experience of what had occurred in Germany under the Nazis, and who were already forming "committees on democracy" in order to foster a nationwide "save democracy" movement. Bateson and Mead attached themselves to one of these groups, a local predecessor of one of the national organizations that emerged a year later. One of the national organizations was the Committee to Defend America by Aiding the Allies (CDAAA); the more radical was the FFF, or Fight for Freedom lobby. Both organizations began their national campaigns against isolationism in 1940 and 1941, respectively. Of the two, the CDAAA shied away from an outright immediate declaration of war against Germany, while from May 1941, the FFF openly stated that the European conflict was an irreconcilable struggle between dictatorship and freedom, and that if dictatorship won in the current arena of conflict, there would be little hope for freedom in the United States. The FFF was convinced that the United States must play a part in the war because the struggle in Europe was truly a world struggle, and America could not avoid this implication. The FFF invited citizens who shared this conviction to join in doing whatever was necessary to insure a Hitler defeat, and in effect, Bateson and Mead had already done so.[2]

Bateson saw his own role in his local campaigning as an attempt to reduce Americans' fear of active engagement and increase emotional involvement in a project that threatened democracy. Their isolationist opponents argued that emotional involvement in supporting democracy was "hysterical," but when Bateson wrote to his mother, outlining the situation he faced, he made the case that democracy depends on "good honest moral indignation and fear too" (Letters, LOC, 23/09/1940). Mead had begun to organize women's groups and advise them on the situation, while Bateson began to pamphleteer against American passivity and defeatism. Mead's involvement with women's groups coincided with the formation of a Women's Division by the FFF, broadening its existing base by providing an outlet through which women could participate. Support for the FFF came from thousands of citizens across America, many of whom sent in donations as small as one dollar. Additional support came from prominent educators, authors and playwrights, clergy, stage and screen actors and actresses, newspapermen, and politicians. Bateson wrote to his mother that, by the end of 1940, their association with committees on democracy had been rewarded through "the growth of thousands and thousands of clubs and committees concerned with 'defending democracy' and many concerned with identifying and evaluating democracy—a new sort of political consciousness" (Letters, LOC, 10/12/1940). Nevertheless, Bateson still worried that a sense of powerlessness to act and intervene in the war still prevailed "except for the communists and other agitators."

Bateson's advocacy led him to contact John Cockcroft at the National Research Council in Ottawa, sending his condolences on the death of Oliver Getty, which Bateson assumed was related to "experimenting with detonating uranium." In another letter of well-intentioned gesture, Bateson offered Cockcroft his own money for the war effort (Letters, LOC, 01/10/1940 and 26/10/1940).[3] His advocacy efforts also led him back to his own writing on the interrelations of dominance and submission that had emerged in his fieldwork among the Iatmul and Bali. It was a theme that stuck deep in Bateson. Powerlessness was another example of the relation between ethos and eidos. He juxtaposed the motif of the Nazi abuses of power with the seeming powerlessness to act in a democracy, and he began to be concerned with those affectual feelings encouraged by propaganda, and which find its "force" in such illusions. In several of his letters from this period, he works through a number of ideas about dominance and submission and finding scientific ways of giving expression to affect and feeling within these processes. To his family and his colleagues, he compares the ways that the Nazis perverted affect to secure dominance with the more nebulous relation between ethos and eidos in the democratic process. One letter argues that it is incorrect to suppose that any question of psychological "cause" arises when people ask themselves what causes people to dominate each other—because dominance in this respect is illusory. The more appropriate way to address the question of dominance is as a sequence of events, and to ask first of all "what makes them think that they are dominating one another? Is power itself a psychological illusion?" Or is dominance a jumble of ways in which the individual can be made to conceptualize his world? Feelings of power or powerlessness arose from sequences of behavior that generated an illusory belief in power, he wrote to his mother when she had challenged him on his understanding of democracy. He informed his mother of campaigns that he and Mead were running in order to show that despite apparent feeling of powerlessness to act throughout American society (except among Communists) there was no lack of power to act and intervene in the war (Letters, LOC, 07/04/1941). In a democracy, he wrote, power and powerlessness is a frame of mind about those at the top having power muted by responsibility, and those at the bottom being powerless, but permitted to raise all kinds of initiatives from below.

Bateson and Mead tried to devise a game that would educate players in the difference between dictatorship and democracy and put it on the market. As well, Bateson sat on an executive committee exploring how to mobilize psychological and sociological resources. He and Mead also began to lobby for a ministry or department of morale, in the expectation that their research would then become available for official use. The department would, in Bateson's thinking, be aimed at relieving the sense of powerlessness, "the continual paralyzing sense of working in the dark," that had seemed to have gripped the United States. The idea was that this department would receive information from various local committees about adjustment, strain, and various types of local strife. It would spotlight desirable solutions—which would then be worked out by local groups.

In March 1941, he felt that he had lost every battle in trying to project his point of view of "imminent danger." All would change as a result of the attack on Pearl Harbor in December of that year. The FFF committees suddenly found their projects for intervention in the war resoundingly successful. Yet their sudden success had a twist in its tail. The issues turned on rumormongering. Public rumors focused on how Pearl Harbor could have happened, or how Pearl Harbor could have been possible without the United States knowing Japanese intentions. From seeming long-term government inaction, the sudden governmental moves toward intervention included government initiatives in support of public informants. The government fostered "rumor clinics"— public discussion groups mostly situated on university campuses. Shortly after Pearl Harbor, rumor clinics tried to control the outbreak of rumormongering on university campuses and then began undertaking an analysis of the origins and effects of rumor circulation. Some of the same personnel as those who had supported committees for democracy became involved, especially in university towns. Professionals in academic departments of social science were directed to support research that would be used to take control of subversive chattering, and other aspects war attitudes. The professionalism of social science, which Bateson had so fervently hoped for in response to an imminent threat to democracy, emerged in quite a different form than he had imagined.

Bateson now probed Canada. The director of public information in Ottawa replied to Bateson's enquiry for possible service as researcher in that country, indicating that he would be a useful intermediary, stating: "We are particularly interested in rumours and loose talk, industrial unrest, and race tension (French-speaking versus English-speaking) . . . and are most interested in the organization which the American Government has set up to deal with the problems of morale in the United States" (Letters, LOC, 25/06/1942). The director wrote to the head of the committee that if Bateson wanted to do research on the French-speaking population in the New England states on this question, he would receive full support from Canada.

As it turned out, the whole rumormongering exercise had a significant role in the development of social science and of social psychology in general, but its significance only became evident after the war. The publication after the war of *Readings in Social Psychology* (1947) devoted a whole section on the psychology of rumor, research the government had generously funded. Both Bateson and Mead have articles in this volume, together with Alex Bavelas, Larry Frank, Clyde Kluckhohn, and Kurt Lewin—all of whom had worked with Bateson—and Ruth Benedict, Eric H. Erikson, John Dollard, and Erich Fromm, who had all worked with Mead. There were also contributions from Mead's joint working group in the 1940s, including Abraham Maslow, Paul Lazarsfeld, a mass communications sociologist, and the two Lynds, both anthropologists writing ethnographies of middle-America. F. C. Bartlett also makes an appearance, along with George Herbert Mead and Jean Piaget. The volume covered a wide spectrum of topics, including socialization of the child, social frustration and aggression, morale, and mass communication and propaganda, all of which were subjects that either Bateson or Mead had engaged in their fieldwork or during the war.

Readings in Social Psychology answered one question, namely what would anthropology look like if its range of enquiry shifted from the study of relatively isolated cultures, many of them under colonial control, to that of a heterogeneous society where the multiple refractions of social interaction in several dimensions required multiple observers. At the same time, it posed a number of pertinent issues about what framework of inquiry should be undertaken in its several applied aspects. The coeditor, Theodor Newcomb, maintained that the role of research in social psychology was to enlarge conceptual understanding of social-psychological imbalances. Though it was not the task of social psychology to advocate the restoration of particular balance in reciprocities, at least the discipline could frame the issues from a general perspective of balanced reciprocity in social action, or, what the psychologist might term restoration of psychological harmony.

NATIONAL CHARACTER

A Committee on National Morale had become formed in the United States in 1942, but the person put in charge was a military man who took on active promotion of morale in one way or another. This led the US Army's Information and Education Division to go through a series of name and organizational changes, each reflecting different aspects of its propaganda function. The Morale Branch was formed in October 1941, before Pearl Harbor. It became the Morale Services Division in June 1942. The formation of the Special Services

Division occurred in November 1942, and then was renamed the Information and Education Division. After the declaration of war, the US Army became open to the introduction of social science professionals. Mead was promptly recruited. Early in 1942, she went to Washington, D.C., to head the National Research Council's Committee on Food Habits. This committee applied anthropological methods to problems of distribution and preparation of food in war-affected countries. A well-known sociologist, Frederick Osborn, friend of both President Roosevelt and Army Chief George Marshall, received a commission, and in the summer of 1943, the Army's Information and Education Division (IED) began planning one of the most massive exercises in social psychology ever undertaken. The cause for which Bateson had vainly campaigned in the United Kingdom three years earlier had at last come to fruition.[4]

Yet military leaders were less concerned about boosting morale at the beginning of hostilities—where they had control over recruitment and training—as they were about the situation that could arise after the war's conclusion. From their point of view, the rapid influx of enlistees would form a potentially restless, dangerous, and uncontrollable group at the end of hostilities—particularly among those stationed overseas. General Marshall in particular had great unease about these future issues because he had lived through a near mutiny of troops at the end of World War I. A near mutiny had come about when the troops were not demobilized quickly enough.

Two of the key issues that emerged were first that servicemen had a negative view of the Allies, particularly the British and French, with whom they had the most contact. Second, they had a negative attitude toward military service generally, since almost half felt they would be more effective working in industries back home. The resulting research became a landmark study in both sociology and social psychology as *The American Soldier* (Stouffer et al. 1949–50). One of its most notable features was the opportunity a sizeable population of researchers had to experiment with new ideas and techniques, especially with automated data and tabulating machines—the forerunners of computers.

Both Bateson and Mead took research funding made available for issues concerning morale and "national character." This latter term, "character," began substituting for their years of research into comparative cultural aspects of "culture and personality" that was the overall framework emerging out of their work on the Balinese and the Iatmul. Mead realized that the overall activity of "personality" emerging in any "culture" could be reinterpreted as the realization of selfhood. Mead's version of culture was dynamic in the way that Ruth Benedict's version of culture—as a static form of configuration or pattern—was not. The latter argued that the life story of the individual is first and foremost an accommodation to custom and tradition, so that "custom is king." It proposed that the tradition into which the individual is born shapes both experience and behaviors from the moment of birth. By the time the individual had grown up and was able to take part in community activities, the community's habits had become his or her habits, its beliefs his or her beliefs, its possibilities his or her possibilities (Benedict 1934).

Mead's national character studies explain ongoing socialization processes diachronically, acknowledging that when culture is perceived through a static perspective, custom and tradition appear to be causal in an overly determinate manner. In Mead's argument, selfhood emerges from "culture" in an overall process of socialization. Unlike the idea of group social instinct or the authoritarian premises of nationality propagated by the Nazis, culture's relation to tradition and behavior was by no means determinate. Mead's study of American national character appeared in *And Keep Your Powder Dry* (Mead 1946). As early as 1940, Mead had begun to use the following definition of character formation when proposing to undertake "national character" studies: "Character Formation will be used to describe the processes by which members of a group sharing a common culture and acting as surrogate for that culture mold the young organism to become a full representative of that culture. [It is] a diachronic concept" (Mead unpublished, 1940). Mead then attached the importance of selfhood in socialization to the development of democratic activity, and presented this dynamic process as "that which we are we fighting for" against the cultural determinism of authoritarian regimes.

Meanwhile, Bateson worked on a major paper about morale and national character that, whether by coincidence or not, mirrored one of the key research concerns of the military, namely the negative views that American servicemen embarking for bases in England had of the British and French. Bateson ran research on small groups with recorded interviews of individuals from different nationality backgrounds, then, taking his descriptive data, he matched it to his own typology of dominance-submission. To this, he added an exhibition-spectatorship, which he regarded as being of a similar universal interactive type (Bateson 2000, 88–106). He was quite pleased with the result, remarking that "some of the [research] stuff comes up quite prettily, especially the differences between America and England" (Letters, LOC, 18/03/1942). He also began to use movies to illustrate what types of differences arise in communication between Americans and British, or Frenchmen and Americans, a study in second-order patterning.

His use of taping interviews and employing film resources led to his comment that, since it requires long reading in a particular culture in order to understand what is meant by being, for example, a "Frenchman," it was essential to find some method of shortening this laborious process so that people concerned with propaganda to the various nations, together with those who are being trained for work in postwar reconstruction overseas, may acquire more rapidly some notion or feeling about those who they will meet. Later, in a lengthy comment on a proposal to establish a school for military personnel to be selected for postwar duties in Japan and the liberated territories of the Malay Peninsular and Indonesia, he stressed the need to include visual material of films and stories specifically to enable students to understand the idea of cultural differences and how such differences can be handled and described. He advised that the most effective way is for the student to understand "fundamental and systemic differences between cultures so that the student is able to develop a point of view about the way in which the differences in cultures are like [they are]." In other words, he calls for a study of comparative understanding of difference in cultures (Letters, LOC, 10/03/1943).[5]

Bateson's view of culture shared Mead's democratic vision but not her attributions of culture merging with selfhood. His view was more in keeping with the idea that culture arises as "a bias" in the habituation of behavior patterns in interpersonal response. His prewar attempts at definition contrasted culture to material substance. About culture and "society" and the differences between these concepts, he explained when writing to Radcliffe-Brown a few years earlier:

> My view of the matter is that "society" is an abstraction of the same order as "a piece of wood" or "an organism"—it is the collective term for a bunch of material units, people [in association]. "Culture" is on the other hand a much more abstract abstraction—of a higher degree of abstraction. Like Society, Culture is a collective term but it applies not to a bunch of material units but to a bunch of behaviour patterns of those units. Thus "culture" is an abstraction of the same order as "mean free path" of the molecules in the lump of wood or "tonus," in the organism. (Letters, LOC, 14/03/1937)

In other words, culture is "meta" to the "bunch of behaviour patterns," but is specific in each culture, in that each culture selects behavior patterns that it integrates into characteristic themes of firmness or flexibility.

His postwar introduction to information theory would give him a more interesting perspective than that of material units and a bunch of behavior patterns. He would discover that there were more subtle levels of difference to take into account, levels that would enable him to make a distinction between aspects of a signal (a material bit) and aspects of a cultural sign (a behavior pattern) and how they both entered into the "bias" of culture. The distinction between the two, signal and sign, was crucial to understanding how the more abstract notion of "signs" allied with "context" shape our cultural perspectives, and more specifically, how signs can be investigated as *gestalts*. He would discover that information exists as material impulses, the transients are called *signals* or "pips," traveling in some arc of a network (viz. the "molecules") while the topological features of the network (behavior patterns) are *signs*. This pattern of "triggering" of response, signs piggybacked on signals, is sharply distinct to the way that behaviorists in psychology approached the issue of stimulus and response. They saw information pips as bundles of energy or as relays acting directly upon the organism's neuromechanisms. Their approach led them to discuss "response" mechanistically, as an automatic response to a direct hit from an information source, which in behaviorist terminology is called a reflex arc. Yet, in information circuits, the input of any stimulus only "triggers" the output but does not directly energize it.

Bateson would then argue that unlike the transmission ratios of *signals*, the transmission ratios of signs are nonquantitative and would have to be written in terms of a ratio in the pattern of the stimulus in relation to a pattern of response, and not directly as a stimulus pattern/an intensity of response, which was the fundamental assumption of all behaviorist theory. He himself had assumed the latter in his research work in Bali on *kris* dancing and trance effects. "In addition, communication between the blocks [of the network] are necessarily non-quantitative [discrete] and have a gestalt form" (Letters, LOC, 21/04/1947). Several years later, he would add to this statement when writing to the famed ethologist Konrad Lorenz, saying quite bluntly: "In principle, the dimension of the physico-chemical universe, mass, length, time and their derivatives of which energy is one, are not simply translated into similar dimensions within the communicational explanation" (MC, 02/05/1966, 876).

PROPAGANDA ON ACTIVE DUTY

Bateson finally achieved clearance for active duty in 1943. At the end of World War II, he received a campaign medal, the Asiatic-Pacific Campaign Service medal, for the work he carried out in his appointment overseas, from February 1944 to September 1945, in the India-Burma Theatre. He spent a year and a half on active service as a training officer, and subsequently, as part of an intelligence group, Morale Operations, for the Office of Strategic Services (OSS). Though he was attached to American forces, he met and coordinated with the British administrators in India and Sri Lanka (Ceylon), and on the India-Burma (Myanmar) border in the Arakan Peninsula, on what he described as "mixed psychological and anthropological intelligence." He described his war effort as being unproductive, engaged in "reasonably useless attempts to apply anthropology in the Far East to propaganda problems" (Letters, LOC, 21/08/1946). It was "interesting" at times, but "a total waste of time so far as any visible effect on planning and policies." He thought he had managed on occasion to help smooth over relations between the Americans and the British Imperial administration in India, and was particularly scathing in his letters about the British bureaucrats attached to the Imperial High Command that he had met in India.

Shortly after the war, he was approached to support a project that would document the work of social scientists in the Office of Strategic Services and other related intelligence organizations. Bateson refused, mainly because much of the documentation surrounding his operations was officially classified as "secret." In fact, Bateson engaged in both "open" and "secret" intelligence work, as was true of other professional anthropologists during this period. There was a wide range of intelligence-gathering operations in World War II that relied on "open" rather than "covert" surveillance, and as many as half of all professional anthropologists in the United States—as was true of Margaret Mead and Ruth Benedict—worked full-time in some war-related government capacity during this period—while another quarter worked on a part-time basis (Price 2008, 1). Professional anthropologists in Great Britain and Australia also joined the war effort in a similar manner, some of them in a military rather than civilian capacity.

The Morale Operations Branch of the OSS was formed in January 1943 to mount propaganda operations of all varieties in enemy territory including "persuasions, penetration and intimidation." It also utilized "black propaganda," which means that the government disseminating it did not acknowledge its existence. The director of OSS, General William Donovan, called propaganda a type of work that was "the modern counterpart of sapping and mining in the siege warfare of former days" (Dear 1996, 17). The OSS had its counterpart in the British Special Operations Group (SOE), which initiated the use of black propaganda early in the war. Morale Operations (MO) of the OSS had a much slower start in using black propaganda than the SOE, partly because the United States entered the war at a later date, and partly because of wartime administration rivalries. Bateson's sphere, in the Far East, was further delayed because the sort of activities that had proved so successful against Germany, such as conducting black propaganda by radio, was next to impossible against the Japanese. Only in the spring of 1945 did the OSS Morale Operations Branch of the South East Asian Command begin producing black propaganda broadcasts, when it used Japanese wavelengths purporting to originate in Sumatra. Later, in the Mariana Islands, the MO beamed daily black propaganda broadcasts to Japan and to Japanese-occupied countries. Meanwhile, the MO engaged in other "dirty tricks," such as conducting disinformation in Japanese-occupied areas and counterfeiting currency, especially the Japanese-Burma one rupee notes.

Bateson reported briefly in articles he wrote after the war on his part in black radio broadcasting to Japan. It consisted in undermining the official radio news of the Japanese broadcast to their troops and occupied territories, by relaying exactly the same information that Japan reported on its own radio bands, except for a doubling of the figures, whether of Japanese casualties or of allied losses. The overall effect, it was hoped, would throw doubt on the veracity of the official Japanese news. There were other propaganda stunts, but as M. C. Bateson records, these seem to have failed (M. C. Bateson 1984, 40–41). As a civilian, Bateson was not required to take part in any military operation, though in August 1945 he volunteered "to penetrate deep into enemy territory in order to attempt the rescue of three agents believed to have escaped after their capture by the Japanese," as the affidavit accompanying his campaign medal stated.

Otherwise, he helped start newssheets, planted news information in established papers, analyzed raw intelligence from intelligence networks of the OSS, and, on occasion, wrote papers analyzing long-term intelligence strategies. Or rather, that is what Bateson stated officially after the war that he was doing. In fact, Bateson played a more active his part, for in January 1945 he was member of the OSS Arakan Field Unit engaged in an MO Operation called "Bittersweet." From September 1943 to May 1944, the Arakan Peninsula had been a heavy combat area between the Japanese and British army units supported by Gurkhas and Indian Army regiments. Both sides had heavy casualties, and the British were forced to withdraw when a heavy monsoon season prohibited the movement of appropriate war supplies, and decimated the troops through disease. The trip of about fifty kilometers, from the safe territory of Cox Bazaar in East Bengal to the front line at Bawli Bazaar (in Northern Arakan State, Burma), had taken British troops four days to make in 1943, but by 1945 the war front had moved beyond Bawli Bazaar, which Bateson reached in the first week of 1945.

The purpose of his trip was to gather as much information as he could about administration and policing in the area now that the Japanese army was in steady retreat, and to begin to publish newsletters about the rapidly unfolding situation. Here, anthropology was of use. There were many ethnic groups—people who could be, and were, used for intelligence purposes, including the Arakanese, the Chittagonian-Moslem refugees, and the Marung mountain tribes. He mentions these in an MO report in the first week of January 1945, a report recounting his trip to Bawli Bazar, writing, "A lot of background material was obtained on Burmese and Arakenese thought and customs in the interviewing process of captured troops" and was attached to the report.

Other issues concerned the viability of allied intelligence networks, and the problem of looking after several thousand refugees, especially the policing and the detention in "cages" of about nine hundred who were known to have collaborated with the Japanese and who "could not be trusted" until their home areas were well inside Allied lines. Finally, there was the attitude of various cultural groups in the newly recaptured areas toward the Allied armies. Japan had constituted its own propaganda branches while it was occupying this territory, and it had made great play on the way in which Japan could liberate the Burmese from colonialism. Bateson noted in his weekly report (at the time a secret report) that the officers undertaking rehabilitation, administration and welfare were "local boys" born in Burma and India, not British born civil servants. They expressed themselves "in sharp opposition to the Imperial Civil Service" on many issues, as was the case also in Malaya, so that cultural attitudes about the Arakanese of the "local boys" in Burma was now of importance. Of his work with "black propaganda," he noted that their commanding officer, a lieutenant colonel born of English parents in India, "would be very shocked at the black aspects of MO (the U.S. Morale Operations Unit)" but the report does not indicate what aspects of the black propaganda might shock him.

THE AMBIVALENCE OF ANTHROPOLOGY IN WAR

Like it or not, Bateson had put himself into the middle of a professional dispute about the moral range of anthropologists in their contributions to wartime situation, a dispute that began to polarize anthropology in the years immediately following the war. From that time until the present day, wartime activity on the part of professional anthropologists has continued to generate fierce controversies within the discipline. There was a very wide range of intelligence-gathering operations in World War II that relied on "open" research on culture in enemy countries, and postwar review of the quality of intelligence gathered in this period suggests that open intelligence gathering, such as analysis of newspapers, pamphlets, propaganda films and the like, yielded better quality information than covert operations. Once covert operations entered into intelligence gathering, and the issue of secrecy along with it, disastrous cases began to occur, cases which Wilensky, in his review of them, termed "information pathology" (Wilensky 1967, 41–74).

At first, anthropologists, like Margaret Mead, who engaged in open intelligence work did not see their professional position as anthropologists in any way compromised. Teaching of anthropology before the war had often meshed with government interests, especially in colonial British territories, and the war tasks seemed an extension of such cooperation. In Britain, Cambridge University was one of the larger disciplinary centers of anthropology, yet the bulk of its students were "colonial cadets." Bateson himself actively disliked the educational connection with the Colonial Office, envisioning anthropology and psychology as a new science with its own subject matter driven by academic rationale. He had to be reminded by the senior members of the Department of Anthropology that, however he may wish anthropology to be taught, the department required students, and the preparation of students for colonial administration, though an applied aspect of anthropology, was a necessary teaching task.

By the end of the war, a number of senior anthropologists had grown thoroughly frustrated with the work assigned to them by the military and political authorities, work which kept on creating considerable disadvantages for the peoples they had worked with prior to the war. Evans-Pritchard, one notable British anthropologist, related to me that he would tell his students with some passion that he was required to disarm the Anuak people of the Sudan, despite the fact that the guns, many of them ancient muskets, play a vital role in Anuak ritual organization and the organization of leadership in their society. Anthropologists who saw policymakers consistently disregard their advice on cultural matters, Evans-Pritchard and Bateson among them, saw little point in future liaisons with government or military bureaucracy and rejected all prospects of an "applied anthropology" devoted to these ends. With the outbreak of the Cold War engulfing many territories in which anthropologists had made ethnographic studies, the standoff between ethnographic research and administrators became more marked.

In November 1944, while engaged in covert intelligence and black propaganda, Bateson wrote a report for the OSS on the topic of the assumed postwar functions of the OSS in the theater of India and South Asia. The heart of his paper presents a familiar topic. Bateson discusses interactive responses in dominance-submission systems (of which imperialism is certainly typical). The report was a reiteration of the position that he had sketched in prior papers before he entered active service. In this particular version, the colonials are, as dominants, highly paternalistic, and the colonized are submissive children, a situation that leads in its worst moments to cumulative "bully-coward" relationships. The interactive dynamics are such that the relative weakness in B only prompts A, "the bully," to undertake greater efforts in dominant behavior that results in further complementary or cowardly behavior, which then prompts the bully to express even greater contempt for the child or coward. In the report, Bateson also ranges over the typology he developed in his papers on "National Character and Morale," linking the bully-coward type of complementary behavior to other forms of interpersonal behavior, such as exhibition-spectatorship.

In his earlier 1941 paper, concerning the issues of American troops being shipped to Britain, he had also tried to answer the puzzling question of how the Allies should proceed with the Germans once Germany had surrendered, so that the Allies did not repeat the mistakes made after Germany had signed the Versailles Treaty following the ending of World War I. The Versailles Treaty put the Allies in a dominant position, but they tired of enforcing their position as dominants. The Germans, on the other end of the interactions, refused to accept a submissive position, which the terms of the Treaty had specified. In this paper, Bateson argues that a future treaty with Germany cannot be organized around simple dominance-submission motifs, and, to avoid these mistakes, other types of interpersonal responses will have to be introduced in order to ameliorate the Allied position of overlordship. His paper asks: What is each of the Allied nations going to do, what dignity is each of the various nations going to give to cultural performance in a defeated Germany? He also questioned how, culturally speaking, the many expressions between the cultures of exhibition-spectatorship become bonded into social interaction of dominance-submission, especially in this case, when it would be the Allies who were giving and the Germans who were receiving food (Bateson 1941). It should be noted that such a mixed field of interaction indeed did occur in the postwar Marshall Plan for aid to Germany.

The Bateson paper to the OSS introduces similar policy considerations. He argues that it would pay the Americans to influence the British to take a more flexible line in order to introduce effective colonial policy in Burma. The imperial system was characterized by two weaknesses, the lack of communication upward from the native population to white administrators, and the British failure to delegate authority downward from colonial managers to the local population. The colonial administration needs to include indigenous colonial officers operating as equals to expatriate officers.[6] Another policy suggestion was for the OSS to take note of the Russians' handling of indigenous people in the Siberia and Arctic regions. The Russians had stimulated native peoples to undertake cultural revival through encouraging dance festivals and other exhibitions of native culture, literature, poetry, and music. Here, dominance and submission was ameliorated by spectatorship, with spectatorship extended to official praise for indigenous achievements in cultural production. Such reversal of paternalist dynamics, he suggested, would ensure that when native revivals did occur in dance, literature, and the like, cultural nativism would not be used as a political weapon against the dominant. Russia had achieved considerable success through this technique of coopting aspects of indigenous culture.

Bateson's elaboration of the differences between the British approach and the American approach in the same memorandum makes cultural sense when considered within the narrower frame of reference of British and American managerial classes. Bateson relates the characteristics of paternalism in the British colonial context back to parental relationships to their children. The claim is that the overbearing paternalism of British rule is related to patterns of relationships established between parents and their children, where dominant parents are at all times the role model for their children; children are expected to listen silently as the parent acts and explains whom the children should watch and imitate. The English family encourages independence in the child by forced separation of the child from parents at an early age and sending him or her away to boarding school—to "stand on their own feet." Stereotypical though this account may be (it certainly did not apply to the British working class), it was true of a great number of the children of colonial administrators in India, the "officer class," who did indeed send their children away from their parents in India to school in Britain at an early age. By implication, this initial pattern of colonial parent-child relations, Bateson argues, eventually repeats itself as an unshakeable pattern in colonial rule between colonial masters and their servants.

The American parenting pattern was different. Many parents in the United States originate in cultures outside the United States, and this immigrant experience had a wide effect on the behavior of American overseas administrators. American parents were more content to encourage independent action on the part of their children, instead of requiring their children to be submissive. The immigrant parent is often in a position to learn from their offspring. Moreover, the American child is encouraged to show off his independence from his parents at an early age in life. At adulthood, this enables the child to show off his or her freedom from dependence on the adult. It also enables the male head of families to parade his wife and children as "exhibit," of their own parental care. Thus, the cycle of paternalism so evident in the British case is remediated, with the American family becoming constituted as a "weaning machine."

By contrast, if British colonials promoted the flowering of indigenous culture, this would certainly mean an about turn in the colonial value system. To admire, as spectators, presentations of indigenous culture that were conventionally regarded as being "primitive," would require radical changes in the perspective of its political leaders. Bateson was suggesting to the United States that an extension in those areas under its control might allow adoption of a sort of "weaning machine approach" to its politically dominant position in the overseas territories it began to administrate after the war—a suggestion that was sufficiently close to the "hearts and mind" approach that it did eventually become US policy. In any event, the CIA kept Bateson's paper in its archives (Price 2008).[7]

The inflexible perspective of the imperial rulers in India is perhaps nowhere better illustrated than when, after the war, the Royal Indian Navy, in February 1946, began rioting in the harbor of Bombay. The British flag commander, Rear-Admiral Godfrey, stated publicly that he would sooner destroy the Indian navy than yield to the mutineers, for he could no longer consider the Indian navy as "proper navy" personnel (Letters, LOC, 21/02/1946).

WAGING PEACE

Several writers have pondered why Bateson regarded his war experiences in such negative terms when he apparently had some success with his work. Price has summarized Bateson's negative response to the following possible motivations: that Bateson began having second thoughts about tactical use of deceitful propaganda; that the information he supplied to the OSS on indigenous cultural issues was not only disregarded, but also used dishonestly; that he discovered that governments would only use information that they wanted to receive no matter what the quality of the research; that the information provider, the social scientist, could exercise little or no control over how the research information could be used; and that Bateson did what he was asked to do within the confines of the limited worldview of his employers, and that later, like other anthropologists who had aided governments in their war effort, he regretted doing so (Price 2008, 8). Bateson's motivations would seem to have included "all of the above" to one degree or another, but in his letters on his return he spoke of a "boring" war, of spending a lot of time reading, and, as he put it, "I brought home with me a profound cynicism about all policymaking folk" (Letters, LOC, 21/08/1946).

He had seen his strong convictions about the role that the social sciences should play in the war dashed one by one. He had been dismayed at the fact that professionals in social science were passed over in the governmental hiring process. He had seen administrators and policy makers who had no fieldwork experience pay no attention to what he regarded as validated ideas that he, Margaret Mead, and others had put together as successful fieldwork methods. Worst of all, he believed that most administrators had little use for qualitative material and often manipulated numerical data in order to disguise their own self-interest in the issues they were promoting. This occurred along with what Bateson subsequently referred to as administrative hubris, namely disproportionate concern to manipulate their world, exercise power, and seek glory.[8] Before leaving for active service, he had published an article expressing the hope that the upwelling of social science in the United States would remain one that was in accord with "science of decency." In other words, his view was that: "morality, spontaneity, initiative, unselfishness and a number of other rather imperfectly defined—but definable—qualities are important ingredients of 'human nature,' and we . . . believe that these ingredients will help us win the peace" (Bateson 1943, 140–142). Not only did he want a theoretical science that could be successfully fitted to "waging the peace," but he also wanted a science that would not treat people as mere puppets for short-term gain, and in the process repeat the Nazi reduction of psychology to ideology and propaganda. Instead, a science of decency must always retain complex qualities that are imperfectly defined that are important ingredients of human nature such as morality, spontaneity, and unselfishness that we are unwilling to educate out of our people.

After the war, as he looked around, he saw that despite the fact that United States had lavished money on social scientific research during the war, the results had been mixed. The heavy governmental endorsement that had accompanied this expansion had resulted in an undesirable shift from qualitative research to the use of quantitative, statistical approaches, as in the case of *The American Soldier*. When Theodor Newcomb skillfully put forth the case for social psychology, he announced that the role of all research in social psychology was to investigate an array of social patterns about "the facts of association," which were imperfectly understood, and present these for education and debate. These patterns of association would contribute a wide array of public information about attitudes that the public held, and introduce more detailed analysis on the psychological imbalances arising from the domination of one pattern of events upon another. However, *Readings* is also a primer on how to employ quantitative data of the type used in the massive surveys undertaken during World War II, Osborn and Stouffer's *The American Soldier* among them. Osborn was very direct in his support for the use of quantitative data. For him, the use of statistics was not merely a new method for undertaking social research, but, as he put it, "for the first time on such a scale, the attempt to direct human behavior was, in part at least, based on scientific evidence" (Osborn 1948–1949, ix).[9]

The war and its aftermath had enormously enlarged the availability of techniques that bureaucrats could use, while claiming that statistics was a guarantor of a scientific approach. The first unanswered question was whether an "applied anthropology" would emerge that was predicated on surveys using quantitative analysis—as was the case for social psychology. Bateson's fear was that the role of classic anthropology, which retained a qualitative appreciation of culture and presented a detailed holistic study of small homogeneous societies, would soon find itself swallowed up in the colonial bureaucracies and become marginalized. Margaret Mead did not share this fear, and on this point Mead and Bateson strongly, even bitterly, disagreed. Bateson's position was that of upholding qualitative anthropological practices, for he had an abhorrence of applied anthropology, of which Margaret Mead was the chief promoter in the postwar United States. Not only did he reject the supposition that quantitative results produced superior evidence in social science, but from his perspective, it was all too easy for an applied anthropology to embrace and enlarge the tentacles of behaviorism in anthropology, to which he and Bartlett were opposed. Bateson raises these themes in a memorandum to Clyde Kluckhohn, the well-known anthropologist, when asked to comment about the development of an anthropology curriculum at Harvard University. He wrote to Kluckhohn:

> The introduction of quantitative methods has tended to paralyze social sciences thinking. . . . A fetish has been made of quantitative methods and whole laboratories have been devoted to solving, with elaborate statistical machinery, problems which had very slight importance. . . . What we have done in the past is to try to borrow quantitative techniques from physics and chemistry and I believe that this was on the whole a naive procedure, the quantitative techniques while valuable whenever they can be applied to social data are in general not well suited to our type of inquiry. . . . If instead of borrowing their techniques, we had borrowed their ways of thinking, we might have made vast strides by now. . . . Especially in our handling of equilibria we get into trouble because our discursive wordy methods of presentation can scarcely handle the differences between stable equilibrium towards which phenomena return when disturbed, and unstable equilibrium away from which the phenomena will move faster and faster. (Letters, LOC, 18/01/1944)

Bateson goes on to argue that if social phenomena were too complex for qualitative handling, this did not mean that formal techniques of natural science need be abandoned, nor did this mean the social scientist had to place sole reliance upon statistics. He suggested that the Harvard anthropology department should look to the wider dimensions of science in which complex techniques, drawn from several sciences, could be used and taught. From biology would come the study of symmetry and the physiological conditions of homeostasis. From ecology, where there are cycles of predator-prey relations that evoke a mathematics of complex oscillations among equilibria conditions (the over-all balance maintained between predator and prey over time). A methodology might emerge that the social sciences had never even contemplated. Specifically, he argued, the curriculum should avoid giving students in the humanities a few lectures on physics, a few lectures on chemistry, and a few lectures on biology, which was the conventional academic way of introducing scientific method.

Instead, there should be lectures on the types of hierarchical order which occur in nature, compared to social hierarchy and self-governing human communities. There would be a comparison of types of equilibria, and how each is linked to a particular form of threshold. Another section of the course would deal with the thorny issue of entropy and the types of order or disorder that random agencies introduce into entropic order. This, in turn, would introduce the notion of "progress in evolution," from an animal perspective, and then "progress," compared along the same lines, in social evolution. Here, the whole issue of learning and adaptive behavior should be introduced. Following this would be an introduction to some formulations and correlations between factors that could be deemed to be measurable. Finally, he would introduce into the course the feedback phenomena in servomechanisms and organisms.

Bateson's proposed syllabus seems too advanced to have been taught in the early 1940s, and perhaps too advanced to teach at undergraduate level even today, but it is an insight into the ways in which he approached his own approach to scientific method in his subsequent writing.

LEAVING MARGARET MEAD

By the time Bateson returned from the war, he himself had undergone several changes of mind about his previous research in Bali. He wanted to revise his and Mead's whole discussion of tensions among groups in Balinese society. He felt that some notions they had put forward about Balinese society were incorrect—specifically that Bali was composed of individuals so entirely devoid of cumulative tensions that the society left no opening for competitive behavior. As he noted in an undated memo in 1946, he also had doubts about the propositions pertaining to the constant downplaying of climax-seeking tendencies in Balinese society, and how they are related to and modified by childhood experience. His revisionist project acknowledged that *Balinese Character* suffered from the "attempt to make our Balinese material immediately available to other social scientists." He reiterated that many theoretical premises remained only "implicit in our arrangement of material and are not explicitly stated as conclusions bearing upon human society in general." This was almost subterfuge for the fact that his whole standpoint had been radically changed, especially by his encounter with cybernetics and the notion of information as "feedback," and he wished to incorporate these two ideas into a new framework for his Bali data.

He wrote to Hutton, another anthropologist:

> The whole state of theory in C[ulture] and P[ersonality, to which he and Mead had been committed] is so vague at present and there are so many things to be ironed out in the field [work] today I should hardly know what the next position would be . . . and if we don't develop some "orthodoxy" pretty soon we shall get into the state that psycho analysis is now in—just a mass of good hunches and a top-hamper of untestable hypotheses. And how does one ever clean up when once a scientific Augean stable has reached this degree of domestic entropy? (Letters, LOC, 20/04/1946)

Bateson's doubts about the value of culture and personality was one of the factors driving their relationship apart, as was Margaret Mead's desire to found a national program for applied anthropology. They lived together for about a year after his return from the war. We learn from Mary Catherine Bateson that her father withdrew from the household in the fall of 1946. Their breakup was cordial enough, and they continued to share work at the Institute for Intercultural Studies and conference meetings, particularly the Macy Conference Meetings on cybernetics that began in the spring of 1946. Together they had formed the Institute for Intercultural Studies in 1944, dedicated to "advancing knowledge of the various peoples and nations of the world, with special attention to those peoples and those aspects of their life which are likely to affect intercultural and international relations." Mead's book royalties and lecture fees went to fund the Institute, as did Ruth Benedict's book royalties after her death. The Institute continued after their respective deaths and closed down only in 2010.

In her autobiography, Mead ties the end of her marriage with Gregory Bateson with their joint shock at the detonation of the atomic bomb: "The atomic bomb exploded over Hiroshima in the summer of 1945. At that point I tore up every page of a book I had nearly finished. Every sentence was out of date. We had entered a new age. My years as a collaborating wife, trying to combine intensive fieldwork and an intense personal life, also came to an end" (Mead 1972, 296).

For both of them, the advent of the bomb was an indication of the dilemmas now confronting the relationship between scientific endeavor and the public at large. Until the bomb, natural science, with physics at its core, was generally assumed to be an agency of progress for humankind—progress that would bring about peace and prosperity in a postwar world. After the bomb, she and Bateson engaged in a number of lobby groups such as the National Committee on Atomic Information. They both joined groups lobbying for the McMahon bill, which proposed to demilitarize the Manhattan atomic project. The bill proposed that nuclear weapon development and nuclear power management would be under civilian, rather than military, control. The bill was passed with little opposition and established the Atomic Energy Commission.

The bomb affected the epistemology of science as a source of truth and human benefit, but the reaction of the couple, perhaps somewhat naively, was to suppose that many natural scientists would now quit their jobs entirely. Mead and Bateson sent bulletins from the Institute declaring that those atomic scientists and other physicists who were aware that their contribution to the war effort had created serious postwar problems might make amends for their current unease by transferring their efforts to the social science fields. Another bulletin supported the formation of the United Nations, arguing that whatever political faults it may have, it was the only possible organization that could tackle the task of eliminating war in the future (ICIS Bulletin, LOC, 14/01/1946 and 13/02/1946).

Bateson published in the *Bulletin of Atomic Scientists* what he called "an anthropological approach" to the pattern of an armaments race (Bateson 1946, 26–28). As mentioned in Chapter 1, Bateson had studied the pattern of arms races before the war, comparing arms races to the underlying pattern of vicious circles demonstrated in "schismogenic" activity among the Iatmul. His concerns were well supported by one of the senior members of the Emergency Committee of Atomic Scientists, Harold Urey, with whom he was in contact. Urey declared that the world must assure that no atomic bombs are made anywhere in the world, nor that any will fall into possession of any government of any kind, since if that were to happen, the world would spend all its days in deadly fear that the bomb would be used.

The Bateson-Mead response, as Donald Worster, the historian of modern environmentalism, tells us, was not untypical. The growth of nuclear weaponry, the secrecy that surrounded its rapid development, the uncertainty of information gathered from the testing sites about the overall effects of nuclear fallout, all "cast doubt upon the entire project of the [human] domination of nature that had been at the heart of modern history." This gave rise to doubts about the moral legitimacy of science, about the tumultuous pace of technology, and about the Enlightenment dream of replacing religious faith by human rationality as the basis of material welfare. Following the first detonation of an atomic warhead, public concern about the testing of atomic weapons, together with steadily increasing nuclear capability of the two superpowers, the United States and the Soviet Union, evoked fears of total destruction of the world through one or the other, or both, making a series of simple mistakes (Worster 1994, 340, 343).

Bateson was to continue to publish his objections throughout his subsequent career, even more decisively when he had public recognition later in his life. In 1979, he resigned from the board of regents of the University of California when the board expressed support for increasing research on nuclear weaponry at the university. His resignation letter repeats the themes of earlier papers, notably that an attempt to initiate "trust" between the United States and the Soviet Union would be much more likely to bring the nuclear arms race to an end than constant escalation of the military capacity of nuclear weapons. His position was justified when the talks in Iceland between Soviet Premier Mikhail Gorbachev and President Ronald Reagan a few years after Bateson's own death did indeed initiate trust between the two countries and so provide a beginning of the end of the Cold War.

LEAVING ANTHROPOLOGY

Leaving Mead led Bateson to leave anthropology. The material gathered from the first few Macy Conferences on Cybernetics encouraged him to suggest how anthropological research might be conducted in relation to cybernetic proposals about dynamics of stability and change. His 1949 article "Bali: The Value System of a Steady State" (Bateson 2000, 107–127) is a sort of valedictory to the older frameworks he had used in anthropology and an introduction to a new phase of Bateson's thinking. This particular article coincided with his taking leave of full-time appointments in anthropology.[10] The article is one of his first in which he supported an information theory–cybernetic approach and in which the cybernetic concept of homeostasis became the ethnographic equivalent of "cultural structure." What he was trying to convey in the Bali article, Bateson told the Guggenheim Foundation, was how fundamentally different the "ethoses" of the Bali and Iatmul were, especially in terms of the effect of change on their social organization. The two cultures had fundamentally different value systems and each achieved stability through different sorts of premises. The Iatmul had a laissez-faire ethos and a schismogenic sequencing of social interaction. The Bali had an ethos where self-maximizing tendencies were to be contained within themselves. Here, a self-correcting process is achieved through constant display of physical balance in social contexts, and by extension of such self-correcting attitudes through emphasis on symbolic balance (MC, 24/04/1947, 585–1a). This contrasted with the highly competitive societies of the capitalist bloc.

The deeper implication was that cybernetics could and should be a means for displacing concepts of mechanical equilibrium as a form of downward reductionism on the one hand, and of the conceptual stranglehold that French sociologist Emile Durkheim had on anthropology of that time. Durkheim's stranglehold in British anthropology was a form of upward reductionism that had occurred largely through the influence of Radcliffe-Brown, Bateson's former mentor, in support of Durkheim's position. A Durkheimian rendering of social structure was always to depict norms and values as a reproducible conglomeration of "social facts as things" passed down from generation to generation. Durkheimian rules were normative and permitted little discussion of social variety within the social incorporation of norms, or of competitive transactions. Information-theoretic rules and cybernetics built their whole set of arguments around responsive communication to "news" and to feedback and feedforward between processes and events.

The subtext was that anthropologists of the Radcliffe-Brown mold had only presented a static, synchronic account of stability in cultural order. Further, Durkheimian rules of explanation merely mimicked the way in which matter-energy relations were expressed in natural science. Cybernetics was superior in that it enabled a presentation of activity in social sciences in a dynamic, processual manner, but with homeostasis permitting oscillation and variety as a fundamental feature of its stability. Unfortunately, Bateson does not express this openly enough in the published article on Bali. Other themes—his analysis of "game theory"—get in the way, and it is to this we turn in the next chapter (MC, 01/02/1949, 879–8c).

3

CYBERNETIC LOOPS

> What I have tried to do is to turn information theory upside down to
> make what the engineers call "redundancy" [coding syntax] but I call
> "pattern" into the primary phenomenon . . . if a pattern is that which,
> when it meets another pattern, creates a third—a sexual characteristic
> exemplified by *moiré* patterns, interference fringes and so on—then it
> should be possible to talk about patterns in the brain whereby patterns
> in the sensed world can be recognized.
> —GREGORY BATESON

Before the war, Gregory Bateson had been the young scientist urging his fellow fieldworkers to go beyond the specificities of ethnographic data and pay more attention to their methodology, specifically incorporating psychology into their discussion of affect (ethos) and modes of memorizing (eidos, ideas, or "logic"), with each being important elements of culture. In his opinion, anthropology would be better off if it explored levels of data by creating typologies or, in the manner of mathematics, devising a matrix of associated features through typologies and the like. Better still, Bateson thought anthropology should move toward becoming a discipline associated with psychology as part of a single behavioral science.

After the war, it is difficult to see Bateson as anything other than a humanist developing the fields of cybernetics and communication theory. He moved against the general trend in the social and behavioral sciences that had developed during wartime toward utilizing quantitative analysis, a trend that ended in marginalizing qualitative methodologies on which so much depended in anthropology. Later, he also grew apprehensive of the way that computer electronic engineers were taking cybernetics into a digital age without due care given to keeping their attention on the analog world (MC, 15/03/1973, 824–2).

He tried to define, defend, and elaborate the way in which descriptive method-ology was deployed, and where possible he encouraged the use of analog logic as a means of explanation.

The war had changed Gregory Bateson, but perhaps more important for Bateson's change in perspective was that the war had changed science. Science was too often "soaked in the habit of instrumental thought," favoring outcomes that maximize singular goals rather than recognizing a variety of values. Both the reductionism and instrumentality of scientific practice makes a natural appeal to instrumental manipulation, which in turn increases the risk of producing "broken cultures" in the future. A broken culture loses all its ability to value those material or spiritual goods that make life meaningful to us all (Bateson 1947).

In the fall of 1946, Bateson wrote a memorandum informing the New York Academy of Sciences of the forthcoming Conference on Teleological Mecha-nisms. The word "teleology" refers to the philosophical doctrine of goals, ends, or final purposes. His teleological mechanism was a mechanism that was in some way able to direct focus on end-states, in this case toward a defined end under varying conditions of its operation or process—from its present to a defined future state. In keeping with the war era, Bateson's example of a teleo-logical mechanism in his memorandum was that of a target-seeking torpedo. The torpedo has a self-correcting mechanism as part of its total causal circuit, which is constituted by the relation that exists between the torpedo and its target.

The process required a governor, an internal program or a number of linked sequences. As a technical operation, this was not a particularly radical idea, for steam locomotives had very simple governors that enabled the locomotive to maintain some stability in its dynamics no matter whether it was going uphill, downhill or just moving away from at a platform at the train station. A setting on the governor controlled the dynamics of motion within some designated limits of the machine's stability, and as engineers began to move from very large mechanical devices such as locomotives, to electromechanical devices such as household thermostats, they came to speak of these mechanisms as "self-correcting."

A torpedo was so common an object that it seemed unlikely to be clas-sified as an object of philosophical speculation, but it is an apt metaphor for self-correction in that its trajectory—once the torpedo is released—permits alterations in the relative position of torpedo and target. In addition, its design incorporated an operating system analogous to elements of physical control, or homeostasis, in all animal bodies. Analogies of this sort are rare in science, yet biologists had known for several generations that animals' bodies had controlled states whose dynamics maintained a stability within defined limits, both relative to one another (for example, heart rate relative to sudden emotional surprise) and to the body as a whole (for example, heart rate relative to physical exertion). This had been termed "homeostasis." The concept of homeostasis originally referred to the interior milieu of the body and its oscillations, but more recently it had become an analogy more commonly used to represent both external as well as internal constraints to survival. The same term is also used to indicate a type of oscillating movement in evolution.

The most important aspect of a teleological mechanism from an epistemological point of view is that it presents a reversal of the way in which causality is defined and controlled in both mechanical and physicochemical systems. Causality in the natural sciences is treated through registering causal sequence from time past to present time, as a sequence of a cause to an effect. But in the case of a moving torpedo honed on a target, causality moves from time present to some expected goal in the future. The immediate question of Bateson's conference was whether the torpedo analogy of homeostasis, or other self-correcting biological examples, would hold; and, if it did, how widespread this inversion of standard explanations of causation might range. The analogy contained another radical implication, for the torpedo was a lineal causal mechanism, which, though it constrained goals or purpose toward a future direction, is designed to do this with maximal efficiencies. In the case of self-correction in homeostasis, the analogy of circular causal anticipative mechanisms certainly applies, but homeostasis introduces the notion of "balance in oscillation," which is not equivalent to the torpedo idea of self-maximizing energies in a linear direction.

The questions that would follow these ideas about self-correction in circular causal mechanisms were many and varied: Had living systems themselves come into being during evolution through types of circularly causal oscillating systems? Or, more speculatively, is evolution circular in its formation, that is, one very large circular homeostatic system? Bateson's memo points out that if the conference was a success in establishing its premises about circular systems, then cybernetics can begin to lay very broad claims against the organization of linear systems, which pervaded the field of natural science. He wrote: "A great part of the philosophy underlying current psychological and social theories may have to be modified and many techniques of experiment and analysis may be made available" (Letters, LOC, 19/03/1946).

The New York conference was not the first on the topic of homeostasis and circular systems. There had been a couple of other conferences before, and the labels for the phenomenon to be discussed varied from conference to conference (Lipset 1980, 179). The list of invited speakers to the first few conferences included luminaries from several disparate disciplines, among them the mathematician and computer pioneer Norbert Wiener. Another famed mathematician attending was John von Neumann, who had contributed in a variety of ways to the development of a general theory of automata, or thinking machines, and had devised mathematical theories of decision making that could be manipulated by such machines. There were Paul Lazarsfeld from sociology; G. E. Hutchinson, soon to become the most recognized ecologist of his time; Kurt Lewin, already the most recognized gestalt theorist; information theorist Claude Shannon; Warren McCulloch from neurology and neuroscience; F. S. C Northrup, an American philosopher; and Lawrence Kubie, a well-known psychiatrist. An Austrian philosopher and engineer, Heinz von Foerster, who was a newly arrived resident of the United States, became the conference secretary. A British luminary in neurology, Ross Ashby, joined them in later sessions. Bateson and Mead were among the regular attendees over the years.

INFORMATION THEORY

In later years, Bateson came to think of Warren McCulloch as the progenitor of much of his own way of thinking about cybernetics and the person who was most responsible for pulling cybernetics away from metaphysics and endorsing it as "experimental epistemology." Yet, at the start of it all, the overwhelming presence was surely that of Norbert Wiener. Wiener was not only the first to work out the mathematics of cybernetics and to give it its name but also the first to present cybernetic applications that applied as much to biological systems as to technical systems, such as a medical prosthesis. Norbert Wiener gave the whole field of enquiry the name of cybernetics, derived from the Greek term for "helmsman" or "steersman." The choice of title reflected several perspectives—teleological, referring to the fact that the conference could revisit the notion of Aristotle's "final purpose," and universal, in that communication linkages of various types are exceedingly common in organisms and communities. Rigorous analysis would lead to approaches to common problems of adaptation, purpose, learning, ego mechanisms, and the like.

During the conferences, in the summer of 1946, Wiener was a visitor to the Bateson summer household in New Hampshire. Mary Catherine Bateson describes him as "smoking smelly cigars and pouring out his latest idea to Larry [Frank] or Margaret [Mead], not much interested in listening to their responses" (M. C. Bateson 1984, 48). Wiener was doubtful at first of the relevance of cybernetics outside the physical realm to the study of society, and expressed such doubts: "I can neither share their [Mead and Bateson's] feeling that this [social] field has the first claim on my attention, nor their hopefulness that sufficient progress can be registered in this direction to have an apprecia-bly therapeutic effect on the present diseases of society" (Wiener, quoted in Lipset 1980, 182).

From Wiener's position, the logico-mathematical rendition of cybernet-ics did not seem to apply in any direct manner to human society, merely to constraints on human behavior in relation to the use of machines in socio-technical settings. And automated work was by no means equivalent to the whole dimension of human activity. Nevertheless, he supported the idea of the conference, "to get together a group of modest size and to hold them together for successive days in all day series of informal papers, discussions, and meals together, until they had the opportunity to thrash out their differences and to make progress in thinking along the same lines."

Wiener was certainly right in questioning the theoretical relevance or ap-plicability of cybernetics and its underlying base, information theory, to social settings. The definitions of cybernetics rested on very abstract scientific gen-eralizations. Wiener had a need to place cybernetics within the universalities of science, and what could be more scientific than to stress the attachment of information to the universally recognized second law of thermodynamics? Once aligned with this law, he could elaborate upon the way that living sys-tems could delay the inevitable movement of thermodynamic energy from an ordered to a homogeneous state. In this manner thermodynamic energy could now be extended to include new ideas about "information." Information was the means through which living systems were able to move against the flow of entropy for a short period of time. In Wiener's terms, information processing by living organisms, including humans, was a constraint on the production of entropy in local thermodynamic systems, and therefore could be considered a sort of "negentropy."

Wiener's definitional approach to feedback as negentropy became revised later by the Shannon-Weaver propositions. In their case, Wiener's negentropy principle could be seen as a special case of a more general method of reasoning about probability and uncertainty—one that does not depend directly on the laws of physics or mechanics (Campbell 1982, 65). The more general source for information theory came from game theory and the notion of stochastic proba-bility. A stochastic series is a sequence of events spread out in time, in which some characteristics may be known in advance, like the price of a share on a stock exchange, but whose events in time are not completely known. In other words, a stochastic series is rather like a series in trial and error. Once trial and error is given a statistically accurate basis for measurement, uncertainty in a stochastic series can be transformed mathematically into an approximate stable regularity. Thus, the closest result in a series of prior information results could be said to bring about an "objective" result in mathematical terms (Campbell 1982, 30, 60).

The Shannon-Weaver version of information remained the standard version in most academic studies of information and communication for the next forty years or so. As Weaver said, the theory was built up from its fundamental elements. From a communications-engineering point of view, any communication system may be reduced to fundamental elements. In telephony, the signal is a varying electric current, and the channel is a wire. In speech, the signal is varying sound pressure, and air is the channel. Frequently, things not intended by the information source are impressed on the signal. The static of radio is one example of this; distortion in telephony is another. All these additions may be called "noise," which is detrimental to communication in a channel.

> The transmitter changes a message into a signal which is sent over the communication channel to the receiver. The receiver is a sort of inverse transmitter, changing the transmitted signal back into a message, and handing this message on to the destination. When I talk to you, my brain is the information source, yours the destination; my vocal system is the transmitter, and your ear with the eighth nerve is the receiver. In the process of transmitting the signal, it is unfortunately characteristic that certain things not intended by the information source are added to the signal. These unwanted additions may be distortions of sound (in telephony, for example), or static (in radio), or distortions in the shape or shading of a picture (television), or errors in transmission (telegraphy or facsimile). All these changes in the signal may be called noise. (Weaver 1966, 17)

The weakness of the Shannon-Weaver approach is that this mathematics of information theory has no reference to meaning. Also, on deeper examination, neither Wiener's cybernetic theory of information nor Shannon-Weaver's probabilistic theory of information gave any reference to "mind," although it would seem to require minds to send any signal down a channel that overcomes noise. This put Bateson into a position in which he accepted and advanced the claims of cybernetics; but, at the same time, could not transfer the mathematical-cum-engineering theory of either Wiener or Shannon-Weaver into a communication theory suitable for the social sciences without undertaking substantial revisions. As he put it in the quotation beginning this chapter (MC, 10/05/1968, 858–144), he needed to turn information theory upside down in order to make pattern rather than coding syntax as the primary phenomenon, and mind grasping pattern to yield meaning as the intended outcome.

Despite individual conflicts and hard argument, the overall aims of the first meetings were a success. By the end of the Macy Conferences, cybernetics had become a widespread metaphor for all linked systems of human-electronic interaction, though the software of 1950s computers was rudimentary by today's standards. At the same time, social science had acquired a new tool to deal with social complexity, especially in industrial organization. Cybernetics grew in salience as more and more electromechanical and electronic equipment was devised having physical servocircuits of feedback embedded within them.

∩ORBERT WIE∩ER

Though Wiener saw no place for cybernetics in social science and Bateson's fierce opposition to cybernetics generalizing its theories of information through promoting technological theories of control, Bateson's friendship with Norbert Wiener was based on other grounds. Wiener never became a "RANDite"—"a soldier of reason," about which we will see more soon. Wiener was a successful person within the military-academic-industrial complex and enjoyed displaying his own sense of importance, yet he did not become caught up in the competitiveness, prestige, and the monetary awards accorded the top ranks of this complex. As a result of his cybernetic ideas, he had considerable success in redesigning antiaircraft artillery, but he had an about-face, in the immediate postwar period, after recognizing the dangers inherent in postwar enchantment with theories and technologies of control. He was horrified by Hiroshima and the prominent role scientists had had in the development of atomic weapons.

Bateson agreed with Wiener's conclusion that the success of the Manhattan Project had resulted in the emergence of "a group of administrative gadget workers [who] now had a new ace in the hole in the struggle for power." Wiener argued that the whole idea of push-button warfare had "an enormous temptation for those who are confident of their power of invention and have a deep distrust of human beings. I have seen such people and have a very good sense of what makes them tick. . . . It is unfortunate in more than one way that the war and the subsequent uneasy peace have brought them to the front" (Wiener, quoted in Noble 1984, 73).

Wiener was disturbed by those men of science who pursued their private ambitions through public administration, especially combined with military ways and means. He felt that they often revealed immature enthusiasms, primitive technical impulses, and a simplistic ideology of automation. Unlike John von Neumann, whose work remained wedded to influence and power, Wiener developed a conscientious policy of noncooperation with the industrial-military complex. He viewed the dominant drives of the military-industrial complex as fundamentally destabilizing to civilian production, and warned one of the national labor federations in the United States of the adverse effects automation was likely to bring to industrial workers. Wiener recognized that the automated control systems on which he had worked reflected human purpose itself. At the same time, this very outcome had to be recognized as its real danger. While it achieved the sort of control human beings themselves wanted to achieve, the achievements of control enhanced the capacities of those who were in power. Wiener even resigned from the National Academy of Sciences in protest over "its official power and exclusiveness and the inherent tendency to injure or stifle independent research" (Heims 1982, 175). He decided to become his own censor by stating that he would not expect to publish any future work that may do damage in the hands of irresponsible militarists. Since the "practical use of guided missiles can only be to kill foreign civilians indiscriminately," he said, "I must take a serious responsibility as to those to who [sic] I disclose my scientific ideas" (Wiener, quoted in Noble 1984, 74).

The major difference between the Wiener and Bateson was their attitude as to "what to do." Wiener was a direct advocate for his position, a public speaker and pamphleteer unafraid to take personal action over the ongoing avalanche of implementation of automation, such as by warning the labor movement of the impending disruption that automation was bound to bring about. Yet, in academic terms, he was reluctant to step out of his own framework of enquiry, for he believed that the nonmaterial concepts of information, such as "mind," "life," and "soul," which were all implicated in the new domain of cybernetics, were unnecessary to scientific description. For him, it was sufficient that the modifications of cybernetics would lead scientists to give a more fully developed description of a physical organism (Heims 1982, 218ff.).

In contrast, Bateson did not engage in this sort of critique of public policy, but had no such qualms about discussing life and mind in cybernetic terms. Even if other cyberneticians wanted to dodge these issues, he would deal with the broader implications of cybernetics, including undertaking criticism of cybernetics itself wherever necessary. A whole range of reformulation had become necessary in the social sciences as a result of cybernetics. Most important, he believed feedback should be recognized as a type of mental event in all communication, in which the flow of information anticipated or developed expectations about the state of a cybernetic system. Feedback adjustment could be interpreted as a form of learning, so the logical implications of feedback would immediately question underlying assumptions of behaviorism. Bateson recognized that cybernetics would challenge inner psychological determinism of the kind that had entered into psychiatry through Sigmund Freud—for example, the supposed repression of libido. It would also modify the notions of "instinct" and "habit" by relating rules of habitual repetition to existing patterns of observed communicative responses, and would thus refute the hard-wired notion of causality in mental activity. But also Bateson saw that the repetitions of anticipation in the perception of living organisms, and their reflexiveness, were qualitatively different patterns from robotic repetitions observed in the digital flows of information in technical systems. Already much of the technical expression of cybernetics was claiming that the iterative information loops in computers passed for analogs of human perception and human thinking.

SOLDIERS OF REASON

Game theory was at the very center of current practices and current theories of psychiatric patient care at the Veterans Administration hospital at the time that Bateson arrived to take up research in the hospital in California. As he was developing some of his first family therapy models, he found his own colleagues were imbued with a vocabulary based on war and adversarial games invoking power tactics and strategies of being "one up" or "one down." Bateson observed how psychiatrists and psychotherapists had absorbed images and models emanating from the war experience and transferred this vocabulary of control to their relationships with patients and clients. He was shocked to find that psychiatrists spoke of their relationship with patients as one of *kriegspeil*, or war games. Bateson's belief, drawn from his understanding of the Richardson process, was that such attitudes "only begat games without end," games that would undermine any therapeutic transformation.

Game theory was tied to the simplifications of mathematical formulae about stages of decision making according to rational and self-maximizing procedures. Game theory sought to reduce life to a number of scenarios and formulas, which then provided knowledge about competition strategies. In short, game theory was not simply a theory about gaming, but provided justification for a whole belief system about interaction and competition through procedures based on rationality. The Shannon-Weaver approach to information theory had become encapsulated in decision-making procedures current in the sphere of politics, in military strategy, and specifically to the way that decision making took place about nuclear missile deployment. While game theory techniques derived from mathematical logic, their suppositions of rationality had also encapsulated an epistemology of selfhood, Bateson believed, one that mimicked competitive self-interest.

A similar worry was that game theory also crossed over into systems analysis. In the early 1950s, the newly founded RAND Corporation came into prominence as a think tank whose research centered on the dilemmas of the United States in its confrontation with the Soviet Union and their respective Allies during the Cold War. When the RAND Corporation began devising strategies for use of nuclear weapons, game theory became a mathematical tool realizing its approach. Its analysts were enamored of computation as a means for determining order as a historian of this period, Alex Abella, commented:

> RANDites posited a new view of human existence—everything of importance in human existence can be traduced into numbers, which serve humanity to keep track of its main driving force, which is self-interest. Since self-interest, according to Kenneth Arrow, is material consumption of goods, then the best kind of government that liberal democracies can promote is that which spurs trained consumption. RANDites therefore established the foundation of a new rationale for Western liberal democracies, one based on the uncontested primacy of the individual consumer—both of goods and politics. (Abella 2000, 51, 92)

Their program was considerably advanced by one RAND member, Kenneth Arrow. His classic book *Social Choice and Individual Values* (1951) concluded that social choice orderings could be derived from individual choice orderings in a simple axiomatic manner. Arrow's paradox, or Arrow's impossibility theorem, presented a supposedly unshakeable argument against the validity of rule through social compact, at least in mathematical terms. Arrow demonstrated that collective group decisions that are rationally based are logically impossible. Therefore, wherever collective rationality prevailed, axioms on social choice necessarily implied that there would be a "dictator." His mathematical theories of human behavior countered Marxist notions of collective will and promoted a value system based on individualism and the economics that accompanied it.

Over the next few decades, Arrow's rational choice theory would become a mainstay of economics and political science, and when, in the 1960s, RANDites moved into the federal government, it would redefine the foundations of public policy in the United States. They assumed that

> self-interest defines all aspects of human activity. Altruism, patriotism, and religion, when taken into consideration at all, would be factored in as variance of selfishness. When applied to corporations, the theory exempted them from any social responsibility other than that owed to their shareholders—as though companies existed in a social vacuum. When applied to government and public officials, it denied qualities such as selflessness

and acting for the public good, viewing public officials as egoistical agents seeking to maximize their own power and budgets, thus equating diligence with self-aggrandizement and good government with veiled tyranny. (Abella 2000, 52)

The other conclusion, stemming from Arrow's work (supplemented by John Nash), was that it was reasonable to assume that collectivities such as nations possessed nicely behaved utility functions like individual RANDites. One step further this assumption implied that given individuals who were rational, and had consistent premises that they sought to maximize for their own selfish benefit, the preferences of utility preferences of nations could be examined in exactly the same manner. By this means, Arrow's rational choice theorems and game theory were incorporated into military analysis, simulating the preferences of the Soviet Union against the preferences of the United States. In the absence of firm intelligence about Stalin's plans, Arrow's paradox joined hands with game theory's prisoner's dilemma as a mathematical basis for simulation exercises probing Soviet intentions.

By the mid-1950s, the researchers at RAND had become the brains behind the Air Force strategy for the deployment of nuclear weapons. Between 1953 and 1955, the American nuclear stockpile rose to nearly two thousand warheads, without counting the additional dozens of hydrogen bombs. Military planners expected annihilation of three-quarters of the population in each of 188 cities in the Soviet Union, with total casualties in excess of 77 million people in the Soviet Union and Eastern Europe if full scale nuclear deployment occurred. To avoid such catastrophic loss of life, RAND researchers put forward the concept of counterforce. Counterforce meant having a reserve nuclear force, or second-strike capability able to respond to surprise attack. It also contemplated using deployment of weapons in graduated attacks, to give the Soviet Union time to come to an agreement before the next wave of nuclear missile was unleashed. Ironically, the concept of counterforce was put forward just about the time that RAND analysts were beginning to lose faith in game theory and the prisoner's dilemma as conceptual tools in making estimates on national defense. Abella calls it a turnaround in their thinking, yet the turnaround did not have immediate consequences. It was only years later that the RANDites themselves ruefully acknowledged the futility of attempting to reduce human behavior to numbers. In the interim: "RANDites [were turning] the concept of human knowledge on its head. Instead of complementing the numerical approach with the softer disciplines of history, social sciences, and anthropology, the economists made them mere adjuncts of the hard sciences, traducing them into numbers, choices, and decision patterns through modeling and systems analysis" (Abella 2000, 51). Their assumptions resulted in a point of view that spread to the whole of the military-industrial complex. "By seeking to numeralize life, to reduce it to a series of equations, formulas, and theorems, Arrow [and others] developed a theory of human behavior that was, improbably, as earthshaking as Marxist dialectics" (Abella 2000, 92). At one stage, Abella notes, RAND brought into its midst a few social scientists to investigate the "human factor" to round out its theories of war, but in the outcome, the dominance of RAND, counting, parsing, and enumeration beat out tale-telling, psychology, and interpretation.

John von Neumann as one of the great mathematicians of his time, was invited to join RAND in 1950, for among von Neumann's noted accomplishments were his work on automata theory and game theory. Though he had only attended one or two of the Macy conferences on cybernetics, his association with cybernetics gave the young discipline credence in the scientific community. Von Neumann joined RAND just at the time that researchers began to expound on their fundamental premise that rational individuals had consistent preferences that they sought to maximize for their own selfish benefit, and that the processes of maximization of benefit were identical in all human beings, and not culturally relative.

Bateson had the greatest difficulty in accepting von Neumann's idealization of game theory (von Neumann and Morgenstern 1953). Not to put too fine a point on the matter, Bateson found that while von Neumann's automata theory might be a shining example of mainstream physics in the mid–twentieth century, game theory posited a world of deterministic dynamics. Ostensibly, the main purpose of game theory is to consider situations where, instead of agents making decisions as reactions to exogenous prices (dead variables), their decisions are strategic reactions to other agents' actions (live variables). An agent is faced with a set of moves he can play and will form a strategy, a best response to his environment, which he will play by, and any game is driven in its dynamics through highly rational, formal, logical criteria.

Bateson supported one aspect of von Neumann's game theory because it gave a good representation of a universal tendency in social phenomena: the tendency of human beings to form coalitions. He even went so far as to praise the premises employed in this aspect of game theory. Yet, since von Neumann and the RANDites supported a social theory that voided Bateson's own understanding of culture and social process, he was committed to invert, turn upside down, the version of selfhood that the RANDites had constructed. In one article, discussing Bali (2000, 107–127), Bateson switches in the middle of his discussion of the relevance of steady state dynamics in Balinese culture, to criticize von Neumann and Morgenstern for applying one-dimensional utility functions to social decision making, and to making equally improbable assumptions that the players in the "game" have both complete information and perfect rationality. He concludes that given these as central component of the way that they present game theory, and that they never discuss how learning the game might change value propositions, their whole statistical exercise is suspect.

A later article by Bateson is even more critical of their simplifying assumptions. The average reader must wonder how Bateson ever arrived at his opening statement that their game theory is perhaps "the most complex and elegant—perhaps also the most significant—theoretical advance that has yet been achieved in the whole field of behavioural science." Nevertheless, in the body of the article, the reader learns that von Neumann's probabilistic frame is achieved at the cost of pursuing "naive premises regarding the biological nature of man and his place in the universe." Bateson tells his readers that their version of probability speaks only to the sequential aspect of time and ignores the durational or experiential aspects of time. In other words, it can tell us only about the direction of change—toward thermodynamic entropy, that is—but can tell us little about duration or stability in a social sense. Real organisms in real environments will resort to particular patterns of search that have little to do with the decision making of individuals as proposed by von Neumann and Morgenstern. Their patterns are only viable in the case of "a searcher with infinite time and in a universe which makes available infinite series of data," a situation that permits "a corresponding restriction in the patterns which can be discovered" (1991a, 141–142).

Bateson's private correspondence expressed even more severe annoyances with von Neumann and his concept of maximizing utility in which each automaton in the game is always busy attempting to maximize access to utility. All economists who resorted to accounting for social activity in terms of utility made a cardinal mistake in supposing that "value" was a quantity, measurable upon a monotone scale. Moreover, their assumption of the players' total orientation to utility gives a static, idealized, representation of decision making. Such assumptions never allow the possibility of individuals learning from their mistakes.

Von Neumann and Morgenstern not only ignored feedback but also ignored how complex is the feedback required in any learning situation. Cybernetics, on the other hand, stresses the continual possibility of error arising from individual action and the need for continual flexibility in adjusting the self to environmental surround. Finally, game theory was not moral, and instead provided justification for a whole belief system about human interaction and social competition in keeping with the overall interests of the military-industrial complex in the United States. Cybernetics provided an alternate conception of control, one that stresses mutual reciprocity and complementarity in coalition formation in a manner that builds on trust, rather than reducing the transactions of the game to noncooperative forms.

The fundamental differences of Bateson's viewpoint to those of the "soldiers of reason," as Abella calls them, are important to understand as Bateson was engaged in a deliberate attempt to counter their epistemology. He would contest not only von Neumann and Kenneth Arrow's finding of self-maximizing utility functions, but also Arrow's subsequent argument that he could provide utility functions for Soviet's collective units as well. As Arrow perceived this latter utility function, he presupposed that group choices could be set up in axiomatic order—that is, that a certain desired result would rank as A, another as B, another as C, etc.—and that eventually a group consensus would emerge. To the contrary, Bateson argued. Human preferences are intransitive, and circular, not axiomatic and self-maximizing. On these grounds, he would also contest Arrow's conclusions that no one could come to an agreement in any decision-making body where at least two players were facing at least three different outcomes, unless someone emerges whose will is imposed on the others (thus supporting an outcome of unilateral top-down leadership). Bateson, as we shall see, supported the notion of heterarchy, derived from cybernetics, and not top-down hierarchy.

COMMUNICATION AS TRANSACTION

Bateson also felt he could not rely upon any of the other definitions of information stemming from the social-psychological theories of his time. Most researchers in social science and related disciplines from the mid-1940s to the mid-1960s—and even later—focused their attention on information as a means of control. Whether it was control of work through group performance, or control of the personality of individuals, or techniques of mass persuasion, the social interaction literature usually reduced information to an aspect of energy or "force." Another set of studies was oriented toward psychic tensions existing between personalities as a type of energy transformation. Typical of the psychic tension school was Harry Stack Sullivan, who provides an interesting contrast to Bateson in that his work covered similar areas of study as Bateson, but with entirely different underlying assumptions. There were only a limited few—Jerome Bruner, Kenneth Boulding, and Erving Goffman among them—who approached information through conceptualization of it as a sign detached either from force or from psychic energy.

"Work" and energy efficiency were key frameworks for assessing group behavior, with a heavy emphasis on behavior relating to commodity production, competition, and planning. Researchers studying groups and organizations in Bateson's time tended to take the expressive aspect of communication messages for granted. The theoretical use of information in social science journals stressed compliance, cohesiveness, and uniformity. Alternatively, expressive events in communication were regarded as being "severely stochastic," that is, a large number of different events follow one another, in sequences that are seemingly random. Or, if they were deemed as being ordered communication, that order could only be analyzed through some type of probability matrix requiring sophisticated mathematics. With these loose assumptions of information as background to the premises of organization in social science, it is easy to see how the RANDites with their supposed objective consideration of rationality, supported through mathematical formulation, could have such formidable influence.

Typically, social psychologists believed that group cohesiveness was an outcome of a shared "force" of individual satisfactions that bound individuals into groups and gave the groups their stability. Often the central concern of the study of leadership in groups was tied closely to this profile of cohesiveness. Such a focus could be justified in case studies of workflow in industrial plants or bureaucracies, but coordination and control of work appeared as the structural themes of relationships in almost any study of organization. In fact, organization was itself defined as the outcome of an individual performing with respect to an "object," which was maybe a tool, but equally might be a symbol or an inanimate object, or another human being (Perrow 1967, 194–208).

A usual method used in estimating cohesiveness of group organization was some form of attitude survey or sociometric study. The latter quantified results from estimates of "attraction" or "repulsion" to various forms of social interaction, with attraction or repulsion estimates drawn from scales devised by the researcher. Usually the scales were linked to forms of "hedonic satisfaction"—that is, the pleasure/pain principle—or other utilitarian criteria (Moreno 1934). Another gambit used was the notion of social power, the basis of social power for any individual being determined as either reward power or coercive power. A third dimension of social power, called "legitimate power," flowed from the group acknowledging the legitimate right of the person or leader in question to prescribe behavior for the other members of the group (French and Raven 1959).

Researchers scaled leadership qualities or social power in groups through one-dimensional responses of individuals, matching these to perfect conformity at one end of the group scale and autonomy or independence or resistance at the other end. Methods often coincided with a checklist approach to leadership criteria, highlighting alternatives of authoritarian, democratic, or laissez-faire leadership, which tended to place all dynamic patterns of leaders' social interaction into preconstructed categories. The semantic meaning of communication in these studies was rarely considered in itself, for communication was presented as a means to some end. Thus, one well-known social psychologist, L. Festinger, defined communication in his study as "instrumental": "By instrumental communication, we mean one in which the reduction of the force to communication depends upon the effect of the communication on the recipient. . . . If the effect has been to change the recipient so that he now agrees more closely with the communication, the force to communicate will be reduced. If the recipient changes in the opposite direction, the force to communicate will be increased" (Festinger 1971, 225).

Interactive personality studies emphasized the notion of hedonic reward and individual response to utility and gain. George Kelly's personal construct theory of personality was a little more complex in that his personal construct theory was informational and anticipative, but still invoked hedonic reward. The individual selects from among those actions that have been most successful in the past in relation to relative pleasure and success. Such selection is associated with primary satisfactions (pleasure/pain) or with more abstract symbolic values associated with other properties that cue rewards. The use of information in these studies lay in the probability of being able to act, such probability correlating with tables registering adequacy of knowledge of how to act in a hedonic manner (Kelly, quoted in Carson 1969, 73).

The study of personality was also based on behavioral sequences pertaining to attainment of high reward-low cost outcomes. Two people came together in interaction, each bringing a repertoire of sets or plans. Interpersonal acts represented prompts or bids to elicit response behaviors within a certain range. In this interpersonal circle, any behavior that was complementary to proffered behavior was designated as being "rewarding." Interpersonal interaction of this sort emphasized balance or congruency. "Balance" in the interaction was, in turn, linked to personality stability or personality change. The predominant assumption was that people would strive to reduce incongruence in their interpersonal relations.

Congruency and incongruence in the interpersonal sphere was complicated by the fact that any person's perception of the behavior of the other was consistent with satisfaction in their own interactional behavior only as they perceived it (Secord and Backman 1964). The self-concept introduced elements of self-reflexiveness into the psychological argument. Even so, observers or researchers then quantified paired enactments of individual behavior. Thibaut and Kelley, for example, made extensive use of interaction-outcome matrices in which the outcome values entering into the matrix refer to the subjective hedonic experiences each person had, or will have, when such interaction takes place. Notably, their interaction-outcome matrices included estimates of the effort or energy expenditure in the interaction (Thibaut and Kelley 1971, 33–41).

The origins of all these theoretical ideas were drawn from the language of transaction in the economic marketplace. And all authors unashamedly acknowledged that their modes of analysis made interpersonal stances no different from any other commodities subject to exchange. Just like any other thing or object of value, the interpersonal marketplace emerged through regulated exchange of satisfaction via negotiated contracts. Normative rules in social behavior were deemed to operate like any other product, as if they were commodities in a marketplace, with the result being that social interaction research befuddled any distinction between human autonomy and variance, and results drawn from interpersonal transactions of an ideal, market-oriented, actor.

There was little mention of feedback in the organizational literature, and when used, the concept of feedback was often put in quotation marks. Sometimes it appeared in studies where the members of the organization negotiated alterations on the nature or sequence of tasks performed by different units. In other words, "feedback" (in quotes) was instrumental in helping coordinate between various actions on objects and, therefore, was a useful means of achieving results through better planning.

REFRAMING COMMUNICATION

As is evident, Bateson's own communication research would owe little to any of these standard approaches at this period of time. One major difference is that while others held, almost universally, to a proposition that communication was a direct product of purposive activity, the action of a single individual engaged with another, Bateson's understanding was that communication was a coupled event between two individuals, in a sequence of other coupled events in which those individuals may engage. All of these coupled events constitute aspects of relationship between those individuals. He insisted that, in the social world, all relationships were aspects of a connective principle stemming outward from the immediate speakers to a network of involved others. What the researcher had to do was to discover the rules of relationship, for the rules give contexts to communication and to intentions in communication. Once achieved, a further examination of communication patterns as response to response enabled an observer to go beyond the actual content of the messages in any communication in order to explore contextual relations. Mathematicians could still consider information as a statistical rarity and retain the mathematical significance of information as "surprise value" or "news," but it became meaningful surprise when a particular pattern of news became—in their terminology—"redundant." In other words, the same pattern appeared in multiple paths of communication events and yielded meaning.

The whole notion of the energy or force of information creating order through interactive exchange needed to be thoroughly revised, Bateson argues. By definition, communicative interrelations arise out of interaction, so that their binding nature results from events that are always relational and from events that have occurred previously between the communicants. Even injunctive commands in the here and now have elements of contextual meaning. Considering communication messages as a set of events occurring solely out of individual brains in the here-and-now moment is a downward reductionism truncating the whole phenomenon of the relationship of communicators. It defines communication in terms of the content of messages. All talk is in some way organized through connective rules, and for the researcher observing communication, not only do connective rules of relationship generate patterns that are meaningful for each individual engaged in communication, but they also permit ways in which an observer can examine the rich tapestry of interaction sequences in family organization.

Another aspect of the same idea is that contexts of communication create the communicator's subjective state in the first place. And if context for an individual engaging in communication requires premises in order to give meaning, and rules about premises trigger patterns of interactive communication, clearly a communicative situation could not be assessed in the same way as an energy or utility transformation. Rather, the communicative situation refers to rule-governed interactions implicating prior feedback. It may not even concern the present communication, but could be formed from a habit linked to a particular type of feedback occurring among entirely "other" communicants. Bateson's perspective becomes a dynamic one. In keeping with his constructivist approach originally developed in *Naven*, different contexts are operative at different points in time, so those in communication implicitly propose, agree or disagree and hold each other accountable for agreements about the interpretation of their interactions (Bredo 1989, 29).

Bateson and his coauthor on systemic psychotherapy, Jurgen Ruesch, had talked about the postwar world as one in which psychological man in the sense of the older notion of faculty psychology was dead, and social man had taken his place. Despite a continuing public interest in one-on-one therapy as practiced in psychiatry, "the age of the [psychological] individual was passed [and] the main stream of events was no longer concerned with the private problems of people" (MC, 20/06/1967, 1241–10c). What was lacking was a connecting link, a logic of in-between, that would enable scientists to connect feedback, person-to-person, person-to-group and group-to-wider social order. Cybernetics, together with information theory, pointed the way. They felt that if they were to model such a connecting link, the focus of connection should not be on the individual person in the group, but upon the circuits of messages as units of study. A circuit of information passes from person through groups to that wider society and returns, undergoing transformation. In this recursive ordering, modern cybernetics would then substitute relations for entities, constraints for linear causality, circuits of information for transfer of energy, messages for facts. In place of single individuals as the foci of attention would emerge the idea of a social network attached to a network of premises, with such premises having differing levels, or heterarchy of relevance.

KURT LEWIN AND GESTALT PSYCHOLOGY

Bateson was drawn to Kurt Lewin and his field theory, largely because Lewin's analysis conceived of the whole psychological situation of the "individual" within a gestalt perspective. Lewin presents an individual as being one part of a whole situation, both personal and environmental. Generally speaking, psychology did not think of the individual in this manner and therefore had little vocabulary to express this unity. Lewin put forward the suggestion of psychological ecology to designate this unity, and his field theory dealt with the "life space" of the individual as a part-whole relationship, the unity of persona and environment. It is the germ of the idea that Bateson was later to include as an aspect of "an ecology of mind."

Bateson had begun to work with Kurt Lewin and his research team shortly after returning to the United States from Bali. Lewin was an acknowledged master of interrelating social interaction with gestalt psychology and perception. Bateson hoped that Lewin's field theory—a combination of gestalt ideas, personality, and interaction—might give him some further resolution of his ideas. He was very impressed with the way in which Lewin placed change as the focal point of his studies, the very thing that was absent in anthropology and that he and Mead had tried but largely failed to accomplish in their study of Balinese society. He could also appreciate Lewin's concept of "life space" in that Lewin conceived of the whole psychological situation of the individual rather than pinning his conclusions to abstract notions of time and space typical in other psychological research. Lewin's field theory concepts traced the creation of "habits" as arising from the life space of the groups to which the individual belonged, and they made the important point that popular psychology, with its insufficient notions of the self and of selfhood, too often attributed personal goals and concepts to the inner region of the person, when these goals and concepts should have been represented as "environment."

There may be different boundaries of "inner" and "environment" for different people, but a "boundary," Lewin argued, should not be viewed as barrier or impermeable closure; instead, as Bateson was to stress later, each can be viewed as a locus of communication and exchange. Thus, the perceptual system of hearing, seeing, and so on, was an intermediate zone between the inner region of the person and environment. Nevertheless he felt that Lewin had not taken sufficient account of cultural differences. He wrote in a brief paper about Lewin's topological approach to life space, "We habitually see a certain time relation between execution and satisfaction of needs . . . but in other cultural systems the relation between need and execution has been differently integrated" (Bateson 1942). Lewin's topological approach to "execution habits" was viable only as a representation of cultural behavior of individuals in the Western world.

The time Bateson spent with Lewin's notion of life space and gestalt resulted in one outstanding paper (Bateson 1947, 123, 127) in which he wrestled with the problem of how to relate psychological habituation, *habit*, to animal learning. The major pieces of evidence available were those studies carried out in psychological laboratories on animals, in the form of experiments on animal learning. He writes to Lewin: "If we have an animal and subject it to a long series of teachings, differing in content but alike in structure (e.g. a series of teachings by Instrumental Avoidance) the animal will gradually learn to learn more quickly in these types of 'context.' Second-order patterning differently orders the typical results of learning experiments, and bridges the gap between the experimental work on simple learning of content and other forms of abstraction which experimental psychology constantly undertakes" (Letters, LOC 06/11/1940).

As Bateson put it in his letter to Lewin referring to his own paper on animal learning (2000, 73–87): "Learning to learn thus fits in as a higher order learning and brings with it all sorts of gestalt concepts but in a measurable [that is, certain] sort of way." He notes that he has attempted to tie up life space material with animal learning so that he can show that the animal acquires a future expectation that the world will always shape itself into such contexts, which is synonymous with saying that the animal is structuring its life space in these terms.

This mode of argument will be considered in much greater detail in the following chapter, but Appendix 1 points out that in a cybernetic world, instead of (linear) causality one deals with restraint; instead of entities, one is concerned with relationship. In the place of objects are "structures," systems and circuits; in the place of facts are "messages"; in the place of transfer of energy is the transmission of information. The difference in terminology supports a new understanding of the relation between system and process. Notions of systemicity in terms of distinct functions or components, that is, as independent pieces hitched together, are abandoned in favor of relatively distinct flows. The flows are integrated with one another as a result of being mutually sustaining. Moreover, they are mutually sustaining in the context of circular processes.

Later chapters will show how relative invariants in the turbulence of ubiquitous change result from patterns formed through cycling of the cycles of change. And insofar as there appear to be distinctive "parts" in these circular processes, this results from inner cybernations holding together more tightly than cybernations between them do. Changes are ubiquitous because cycles continually compensate for their own deviations and departures from their relatively invariant patterns of cycling, and these deviations always produce change in their ongoing circularity. The language of balance becomes that of boundaries and separation, connectivity and inclusion, and in so doing it turns many conventional assumptions of distinction and integration upside down.

4

WHY WE SEE IN OUTLINES

By inserting cybernetic notions of feedback, learning, second-order communication, and circular causation into all living systems, Bateson felt that the ubiquity of these notions would create a better understanding of how the seeming ephemera of information affected all aspects of organization in life. In biology, mindful communication would replace existing notions of information theory's mindless materialism, and would, in the long term, enlarge an understanding of the association of human beings with their natural world. Meanwhile his objections to behaviorism, logical positivism, and positivist sociology, gained appeal in the countercultural movement of the 1960s, for it attacked that which the counterculturalists believed underlay the bland authoritarianism of the times. Theodore Roszak spoke for the counterculture when he argued that the existing canons of rationality, gave too much intellectual support to the activity of the military-industrial complex, and too much authority to the state. Attacking the canons of rationality should be the central focus of the counterculture's resistance (Roszak 1969). The myth of the rationality of the military-industrial complex lay in its supposed "objective consciousness," which Kenneth Arrow and his colleague at the RAND had so avidly perpetrated. The counterculture should reject objective consciousness, with its false rhetoric of utility and self-interestedness, and challenge its top-down hierarchies of command and control.

Both Roszak and Charles Reich (Reich 1971) wrote about the effects of a Cold War society that had become drowned in images of mechanical ways of thinking, crowned by the lurking presence of the new giant computers. In those days the size of computers covered the floor space of whole rooms. Their giantism reminded everyone of the risk of their use to destroy a major part of humanity, especially when computer-guided intercontinental ballistic missiles (ICBMs) became part of the platform for a strike with nuclear weapons. The nightmare of the 1950s and 1960s was an arms race with missiles. The threats became even more alarming with the launch of Sputnik in 1957, when the Soviet Union showed it was devising a space program that would eventually use guided missiles able to hit populations anywhere in the world. Suddenly the United States felt that it was in catch-up mode. It launched its own version of space orbiting rocketry in October 1959. Roszak probed perceptions of giantism in the technical order, a deeper level of abstraction than daily fears of war, and the gestalt of a vast electromechanical order in which people felt trapped.

Allen Ginsberg, the well-known poet, summarized the underlying feeling of the years between the late 1940s and the early 1960s when he said that the missile race had brought about "a definitive shrinking of sensation or sensory experience, and a definite mechanical disorder of mentality that led to the cold war. This desensitization had begun a compartmentalization of the mind and the heart, the cutting off of the head from the body, the robotization of mentality" (quoted in Turner 2006, 62).

The countering of robotization became the focal point of the counterculture's activity in its early years. Only later did the counterculture engage in political protest. As we have seen, Bateson had an understanding of the phenomenon of vicious circles, how they formed, and how they escalated. He knew that neither one side nor the other could control the events to which they were party. Their interaction was nonlinear and could rapidly oscillate out of control. Only through change in the segmentation of their joint interaction, and only through a reversal or inversion of their interaction, could the incremental patterns of feedforward in the arms race, and the vicious circle of the Richardson process, be diverted or stopped.

How much the counterculture knew of Gregory Bateson's schismogenic approach is not certain, but cybernetics' discussion of nonlinear patterning was especially appealing. It emerged in science fiction at this time and prompted imaginative alternatives to the present conditions of giantism. Not only was this literary discussion in science fiction a discourse about the top-down flows of power and rationality, but it also provided in fascinating detail the imagined flexibilities of an informational world in contrast to the world of mechanism.

For all that, Bateson's work remained difficult to read, and until 1972–73, when *Steps to an Ecology of Mind* was published, it was difficult to find. Even his students or colleagues who had read some of his work in anthropology or family therapy wished that they had had the advantage of being able to read an overall understanding of his scattered writings. Bateson's first attempts to produce a volume of this sort had run into difficulty when the publisher he wanted turned down his own proposal in favor of another volume in which other authors reproduced Bateson's seminars about his notions of communication (Harries-Jones 1995, 27). When *Steps* appeared, it was evident how inter-personal interaction, as a communicative event, was linked to his prior study of culture and his ongoing study of ecology.

Bateson tried to ensure that his arguments would not be mistaken as a type of worldview. His writing placed an emphasis on perception rather than cognitive rationalism, suggesting a holistic appreciation of life—if human life survived the nuclear missile arms race. But, unlike the leading writers of the counterculture era, he did not give way to their idea of the evolution of new forms of consciousness. One such writer who particularly drew his ire was Teilhard de Chardin, whom Bateson believed confused both evolution and cognition with false notions of a divine creation. He totally rejected de Chardin's Catholic ideas of a supposed new phase in evolution, the "noosphere," appearing through an increase in organismic mental capacity over eons of time. Among other things, Bateson believed that de Chardin confounded change in living systems with technological advance.

COUNTERCULTURALISTS

Charles Reich's best-selling ode to the counterculture, *The Greening of America*, presented the counterculture in terms of a new consciousness of personal freedom and egalitarianism. He mixed sociological analysis with praise for rock music, cannabis, and blue jeans, arguing that all these fashions embodied a fundamental cognitive shift, which he termed "Consciousness III." His Consciousness III followed from "Consciousness II," a viewpoint of a society that featured meritocracy and the growth of large institutions, as in the society that had flowered in the years of the New Deal, World War II, and the 1950s. Before that was "Consciousness I," which according to Reich was the worldview of rural farmers and small businesspeople in nineteenth-century America (Reich 1971).

In contrast to Reich, Bateson believed that a philosophical or operational search for a distinctive, located human consciousness was a red herring and a waste of time. Whatever "Consciousness" lay in the brain, there was a larger unit, a "semi-impermeable" linkage between consciousness and the remainder of "mind," both in the body and through external communication with the immediate environment. There was a much larger totality to take into account than the "Consciousness" that Reich presented, for only a certain amount of information about what is happening in this larger part of the mind comes to be relayed to what "kids and mystics" called their notion of "consciousness," or, a concretized perceptible consciousness. In other words, part of what he was later to call the unconscious or self-organizing ecology of mind became excluded when the kids and mystics centered all their attention on Consciousness with a capital "C." Instead, what gets to perceptible consciousness is selected, a systematic—but not random—sampling of the rest of mind. A screening process ensures that the conscious "I" (ego) sees an unconsciously edited version of a small percentage of what the retina senses. There is always a wider dimension to the unseen in any conscious recollection of what an individual reports as an eyewitness (Bateson 2000, 432–445).

The Charles Reich–like beatnik and hippie versions of consciousness were a symptom that the "dropout" generation no longer trusted the rhetoric of patriotism behind the calls for military action and support for the nuclear defense establishment. Nor did they favor the tools of management and control, which in their understanding stretched from government to the educational system, and downward to the unexciting curricula of the university. We might expect that the dropouts would clearly mark their presence. They displayed different costume, different language, and a different set of orienting ideas to accompany the bodily costume. The dropouts may have become notorious for setting up free-love communal organization and experimenting with drugs, marijuana and LSD. Yet their back-to-the land movement was largely self-subsistent and was a serious attempt to live outside the orbit of a society framed through property transactions.

Their collaborative approaches toward work were equally serious; through establishing rural communes, they hoped they would create a path for new forms of ownership, since they believed the ownership relation had become smothered by the industrial-bureaucratic-conformist society in which they had been reared. So, too, their attempts to create alternative forms of sexual partnering and sexual intimacy, they hoped, would create a path for more equal relationships between men and women. Their notions of change in "Consciousness" supported this attempt to create a different discourse among themselves. It led many of them to becoming interested in the transcendental frameworks of Eastern religions and Buddhism, with its associated physical and mental discipline of yoga. Although the term *yoga* was typically associated with Hatha Yoga and its postures, or forms of exercise, the translations of the Sanskrit word "yoga" included "joining," "uniting," "union," and "conjunction"—terms that embodied notions of relatedness, shared consciousness, and social networking.

They identified "collectivity" as a series of friendship ties and peer-to-peer linkages, which acted as a means for coordinating decisions in the absence of formal institutional ties. They had made a decision to opt out of society's norms because they believed that the normative was a dangerous threat to their survival. The new order that they wished to bring about would dissolve top-down hierarchies of command and control in favor of some form of network, social networks as we would call them today—a bottom-up society.

The drug side of the counterculture did not seem to worry Bateson much. He offered himself to be a subject for LSD experimentation, but expresses disappointment at its outcome. He remarked to the organizer of the experiment, Joe Adams: "This stuff is all very well. It's very pretty, but it's trivial. . . . It's like the patterns of breaking waves or glass. What I see is only the planes of fracture, not the stuff itself . . . and *what that stuff* is, Joe, is quite another question" (1987, 70). Bateson was not convinced that drugs like LSD explained anything about "what sorts of thought" are the stuff from which spontaneity or creativity is made. Bateson's reaction was that experimentation with drugs like LSD would lead to trivial conclusions about any question of how human spontaneity and novelty occurred.

Bateson had a low regard for the intellectual quality of the ideas of beats and hippiedom in general. He worried about their anti-intellectualism when questions of mind and self preoccupied their discourse: "the anti-intellectual wave which seems to be flooding in [through] encounter groups, hippies, much of modern art, student activists and probably Richard Nixon" (MC, 11/09/1970, 771–6). His attitude did not stop him from meeting with, or even working with, some of the leading figures. Both Ginsberg and another Beat Generation writer, Alan Watts, author of the books *Zen Buddhism* and *Taoism*, worked directly with Bateson at one time or another. Nor did his discomfort for the anti-intellectual wave stop him from having essays appear in such publications as *The Coevolution Quarterly*, one of the main journals that catered to the counterculture of the mid-1970s to the mid-1980s. His perceived association with the dropout generation became enhanced through appearances in Stewart Brand's *Whole Earth Catalogue*, while Brand's *Two Cybernetic Frontiers* (1974) had a section on Gregory Bateson.

Bateson arrived as visiting senior lecturer at Kresge College, University of California, Santa Cruz, in 1972, the year that *Steps to an Ecology of Mind* became published. Soon after he gained unusual attention within the University's student population through his teaching. Students began to notice that instead of having graduate students as teaching assistants in his courses, he had professors well known in their own disciplines, including the provost of Kresge, dropping into his course as his TAs. This was unprecedented, and students were quick to pick up on that practice. Bateson had come to Kresge College at UCSC at a time of encounter groups, self-help, the growth of new forms of computing including personal computers, multiculturalism, and environmentalism. His former marriage with Margaret Mead identified him with anthropology and feminism, while his cybernetics seemed to the students to put him at the center of other trends. As time moved on, former research members of his team at Palo Alto turned into authors such as Paul Watzlawick, Jay Haley, Virginia Satir, and John Weakland, all of whom cited Bateson as a major influence of their approaches to psychotherapy. And Bateson became the central subject of a book, when John Brockman probed the deeper levels of Bateson's epistemology in *About Bateson* (1977). At this time, neurolinguistic programming (NLP) was also growing in popularity and brought with it a form of cybernetic epistemology that merged academic credentials with popular styles of teaching designed to change individual perception; Bateson lent his support.

Bateson's position as a statewide and national figure grew quickly after the governor of California, Edmund G. Brown, appointed him to the board of regents at the University of California. During the 1970s, Jim Eicher, who was associated with NLP, became one of Bateson's undergraduate TAs. He speaks of Bateson at this time as not playing the role of guru or counterculture hero, as many others did during that eventful era from the 1960s to the mid-1970s. While Bateson recognized that he had an unusual perception of the world around him, he used his occasions for public speaking or teaching short courses outside the university mainly as a means to put together sufficient money to get time off, for finishing his writing of *Mind and Nature* (1979). As Bateson's evident successes grew in number, Eicher found that he remained "always approachable, kind, thoughtful, and completely (probably to his disadvantage) nonhierarchical in his personal thinking." However, he told me, Gregory Bateson was sometimes frustrated by people not understanding his point of view. More recently, Eicher has written his own book, which reflects on his own interaction with Bateson's ideas.

PAPERBACK WRITERS

Unless any social movement can find avenues for disseminating new ideas and a means for sharing ideas with others, the movement's discourse will inevitably drop away out of mind and out of sight. The countercultural movement required a physical place or other focus for those committed to active participation, and the paperback bookstore provided both location and means of dissemination—a dollar or so for a paperback book, less if you picked it up secondhand. The paperback was as influential at this time as the tape recorder was to become in the late 1970s in providing a means for disseminating ideas in the Iranian revolution, and as the Internet and cell phone integration in recent years fostered a shift toward electoral reform in the Middle East. Nowhere was the paperback bookstore more evident than in northern California. Ginsberg, along with other members of the War Resisters League, circulated around the new paperback bookstores, City Lights, Cody's, and Kepler's, which had recently opened in San Francisco, Berkeley, and Stanford. The League was a peace organization, which, in the 1950s, had local branches in California (the national organization was to play an important part in the anti-Vietnam demonstrations in the mid-1960s). City Lights remains today as an icon of that period.

By 1953, the *Atlantic Monthly* had reported that paperback stores in the United States had become a highly competitive mix of serious literature and trash, of self-help and pseudo-science, of sex and inspiration. Paperback bookstores had been derided in the immediate postwar period as being centers for the purchase of dime novels of insignificant literary merit. Over the next couple of decades, these stores proved to be the social center for artists, poets, songwriters, folk singers, and bands such as the Grateful Dead, all of whom helped to define the beat and hippie generation in California. The same bookstores drew university students like Stewart Brand.

Steps to an Ecology of Mind appeared as a paperback and certainly sold well for a book of high intellectual content. At least part of its success was due to the way in which the actual play of ideas within it was accompanied by a style that matched that play. The metalogues that open *Steps to an Ecology of Mind* were an approach to how a discussion of ideas could and ought to be undertaken in a nonlinear manner. Bateson broached the subject of "illusion" and "reality," showing how falsified patterns of communication, leading to lack of trust and meaninglessness, can intensify unresolved paradox. If one believes that one's own political leaders are delusional—as did those who uncovered deliberate misrepresentations of the Vietnam war by the military and the politicians—then there is a disconnect in institutional life between what is real and what is imaginary. The theme of disconnect and paradox appears in Bateson's sympathetic account of schizophrenics who are unable to distinguish voices they hear emanating outside themselves from their own thought processes. Bateson's arguments that mind is one level more abstract than substance and mechanism, invoking both aesthetic and ethical relations, were entirely in keeping with the counterculturalists' own judgment about the delusional rationality of those running the country. Joining one's own spiritual life with Zen Buddhism or other mystical forms of East Asian religion was perfectly rational in the light of such political delusion.

Toward the end of his life, Bateson became reconciled with those mystical transcendentalists of the counterculture, especially those at Esalen who had seemed to take his own writings to heart:

I have, after all, chosen to live at Esalen [in California] in the midst of the counterculture, with its incantations, its astrological searching for truth, its divination by yarrow root, its herbal medicines, its diet, its yoga, and all the rest. My friends here love me and I love them, and I discover more and more that I cannot live anywhere else. I am appalled by my scientific colleagues, and while I disbelieve almost everything that is believed by the counterculture, I find it more comfortable to live with that disbelief than with the dehumanizing disgust and horror that conventional occidental themes and way of life inspire in me. They are so successful and their beliefs are so heartless. (Bateson 1987, 52)

In the interim, the communalists of the counterculture had a change of vision, which in turn had a profound effect on their social movement. This occurred especially among those who did not want to drop out forever and who wanted to return to the disciplines of employment. Many of them, avid science fiction readers and having more than passing interest in cybernetics, had an inclination to join the ongoing technical revolution in computers and electronic data processing. Publications of the counterculture era shifted their perspective in order to incorporate the technical advances in personal computers, which seemed to promise the means through which a decentralized, networking society could emerge.

After Bateson died, Brand helped build and maintain a deep association between the ongoing development of computers and the migration of the New Communalists back into society "by continuously depicting the desktop computer as a 'personal' technology in the New Communalist sense, and by linking computer hacking to New Communalist attempts at Comprehensive Design" (Turner 2006, 246). The transition from counterculturalist to "wired" is best illustrated in the form of the executive editor of the influential journal by that name, Kevin Kelly, who was a former editor of *Whole Earth Catalog*.

PERCEPTION

Bateson had been working on a set of information theories that were not merely an alternative to the conventional theories of rationality and mechanism, but that had at the very base of their epistemology an inverted, or "upside-down," view of conventional ideas in social science, natural science, and even cybernetics. Conventional studies relied too much upon materialist models of "knowing," and became misleading when they depicted a "hard-wired" view of the brain in which knowledge supposedly derived from processes that integrated separated bits of data, much like a computer loop. His alternative view turned the focus away from the study of cognitive functions in mind to evidence about the correlation of perception to memory recall, where perception arises out of the experiences of communication. Perception was a mapping of a complex territory; but while the hardwired orientation of cognitivists generally assumes that thought in the human brain is a representation of the territory it surveys, perception could make no such claim. Instead perception acknowledges that differences exist between the perceptual map and the territory, as the semanticist, Alfred Korzybski, had maintained in his distinction that "the map [of perception or of cognition] is not the territory." The indeterminacy of Korzybski's view of perception, which became part of Bateson's writing in the postwar period, helps explain why he entirely rejected those who pursued the idea of Consciousness with a capital "C."

While indeterminancy is usually regarded as a weakness in natural science, Bateson discounted this objection, and saw that the very indeterminacy of perception could lead him to an understanding of how patterns relate to meaning. Perception gives us an ability to probe distance and gain the time needed for intelligent reactions to ongoing events. Perception applies only to the immediate future, generally less than one second ahead in time. Its time frame is one of anticipation, while in conceptual events the present is read from the past in order to enable prediction and planning for the future. This difference in time frames supports a distinction between percept, and conscious conceptual activity. Though perception is not traditionally thought of as being an intelligent activity, it can generate hypotheses. As *The Oxford Companion to Mind* points out:

> Perception is remarkably fast, and needs to be, because unexpected events do happen and they can be dangerous. In this respect, perception differs from science—and differs from our conceptual understanding. Indeed it may be said that we have perceptual hypotheses and conceptual hypotheses of the world (and of ourselves), which are different, and which work upon different time scales. The factor of differing time-scales provides a clue to why perceptual and conceptual hypotheses differ, and why they may conflict. For it would be impossible to access the entire corpus of our knowledge for each perception, occurring in a fraction of a second. (Gregory 1987, 610)

Perception gives us knowledge of the states of our bodies in relation to objects and events in an environment. And it is perception of the immediate future that allows us to survive in a threatening world. Yet people in the Western world, influenced by studies of cognition, discount the unconscious nature of the processes of perception. It is usually a surprise to find out—as when reading

Bateson—just how little we are conscious of this day-by-day, second-by-second activity of perception. Most people assume in a very real and conscious manner that they see what they look at. As Bateson wrote,

> when we say that what we see, feel, taste, hear some external phenomenon, or even some internal event, a pain or muscular tension—our ordinary syntax for saying this is epistemologically confusing. What I see when I look at you is, in fact, my image of you . . . these images are seemingly projected out into the external world but . . . "the map is not the territory," and what I see is my map of a (partly hypothetical) territory out there. (1991a, 204)

The total unconsciousness of the processes of perception occurs because our mental "machinery"—or what we know about it—provides only news of its products of perception, and no news about the processes of perception. For Bateson, the unconsciousness or nonconsciousness of the processes in our mental states was never the same problem that it seemed to be in standard scientific approaches. What was problematic was that modern science seemed either to ignore all this evidence about the importance of the unconscious in forming our perceptions, or to offer mystical accounts of these unknown perceptual processes.

From his background in biology, he knew that in no animal does a cortex exist without lower parts of a central nervous system. And in no human individual does any secondary system of consciousness exist that is not mediated or accompanied by a primary system (MC, 01/02/1949, 879–8c). When it comes to action sequences such as "play," it becomes irrelevant to invoke criteria that which distinguish consciousness from unconsciousness. A person may see himself as "playing" while unconsciously he is "not playing" and vice versa— he may consciously think he is "not playing" while unconsciously he thinks he is playing (Bateson 1951, 14). In other words, consciousness, unconsciousness, and thought are bound up with perception, and in the process alter any ability to grasp concrete "data" unmodified by perception. He held that there are no separate divisions within the brain dealing with the reality of consciousness on the one hand, and the fleeting aspects of perception, or the like-dislikes of affective feeling, on the other.

Perception begins with the actual circumstances of vision, and in normal circumstances the retina receives images upside down, and the "inside" brain has to interpret these images, a process that is entirely unconscious. When he participated in sophisticated perceptual adaptation experiments run by Adelbert Ames, Bateson learned that some aspects of the topsy-turviness of perception can be consciously perceived. Ames induced Bateson to join one or two of his experiments on inversion of perceptual images and their reversion of inversion. Bateson, as experimental subject, wore spectacles designed to invert our perceptual images, thus giving him the opportunity to examine the changes that come about through this inversion. This procedure gave some insight into perceptual inversion. The inverted image could be successfully corrected after a period of time when the inversion spectacles were removed. The inversion-reversion process produced interesting information about perceptual change, but, at the same time, shook Bateson's own belief in the fidelity of his own normal perceptions (Bateson 2000, 135, 469).

OUTLINES AND REVERSE HIERARCHY THEORY

From early fieldwork days, Bateson held that patterns of perception constitute high-level macroscopic vantage points that enable us to ignore much of the microcosmic world that materialist science investigates. Thus:

> physics and chemistry theory invokes all the time abstractions smaller than observable objects—while we invoke abstractions larger than the observable objects. If instead of being human individuals in a "society" we were molecules in a lump of wood, to explain our behaviour we should have to invoke the real existence of the lump which would be beyond our ken. Those of us with mystical tendencies would then devote a great many words to discussing whether the lump had real personality in the sense in which we attributed personality to ourselves—and so on. (Letters, LOC, 14/03/1937)

In the macroscopic world of perception, ideas give rise to other ideas, symbolic events give reminders of other symbolic events, and we create analogies. Bateson's whole approach to the perception of shape and form is expressive of this capacity to abstract, classify, and generalize.

A macroscopic world of perception is a characteristic of all living systems and has to do with why human beings see in outlines.

Bateson's metalogue "Why Do Things Have Outlines" (2000, 27–32) plays with the pun that human beings draw outlines to be clear in their discussions and conversations, but that human beings can never really see an outline of a conversation until the conversation is over. "You can never see it (the outline) while you are in the middle of it," and by the time the conversation is over, when you can see the outline, it does not matter whether the conversation was clear or not. But then, human beings are not like machines and his punning is intended to support the conclusion that pattern, not reason, is the underlying phenomena of perception.

Recent research has revealed that infants, despite their physical limitations and lack of experience of the world, have impressive perceptual skills. More specifically, infants regard each speech sound as an analogue of specific arm movements or gestures. Infants are sensitized to speech sounds through generating neural motor programs, which match motor programs of arm postures and movements. This makes it possible for very young infants to match speech sounds to sucking or head turning postures. Typically, infants are able to understand about 150 words and can successfully produce 50 words by eighteen months of age. Eye-related motor control improves dramatically over the first few months of life, and by four months they have already developed a perception for objects by reaching for them; by six months, infants can estimate an object's distance, orientation, and size. Their major means of categorizing one object from another is through shape rather than size or texture. The dominance of visual shape at this age of life suggests a facility for learning categories of objects through recognizing difference in shape. These links between perception and action are well grounded through research, but there still remains a mystery about the interface that associates nonconscious perception of language with conscious social cognition (Allott 2005).

These findings would have pleased Bateson, for they support his key notion that the perception of difference creates difference in patterns perceived and relates to information of a different order. In a second draft of *Angels Fear*, Bateson discusses perception of movement in this way:

> the movement of image relative to end organs on the retina has the function [in human beings] of scanning, a function of turning a difference which was in a sense static in the image, into an event, a change. And if you in one way or another prevent the micromystagmus, the eye ceases to be able to see. . . . In some men the afferent impulse—for that is what we are talking about, may be news of an event outside of O[bserver] or it may be news of a difference between one part and another of the environment of O. And to make it more difficult of course, it is absolutely essential that O know the difference (that word again but now in a different sense) between those impulses which come in triggered by an external event and those impulses which come in triggered by that synthetic event created by the end organ itself. In other words, if I move my head and get that extra information by parallax, I have to be able to sometime ("I" whatever that is) to distinguish between these two ways of seeing. The knowledge of depth obtained in parallax, the increment of information, results in information of a different order.

In motor theory, the visual perception of objects is a process through which the infant's stream of visual experience extracts neural representations of the shapes of concrete objects and produces ambient language. The eye is always active and scans objects by a rapid succession of movements. Motor commands for the eye muscles produce movement of the eye up or down, from side to side and obliquely. According to modern motor theory, there is motor equivalence in the flow, so that the direction may move from perception to gesture or gesture to perception, but in both cases speech and gesture arise as interacting elements of a single system. As Allott writes, "every gesture structured by a perceived object or action can be redirected to produce an equivalent articulatory action. Motor equivalence can function between speech and perception and between perception and motor action because perception is also a motor activity" (2005, 15).

After his study of Lewin, Bateson was also committed to gestalt theory. The recent reverse hierarchy theory (RHT) substantiates gestalt's version of perception, namely that the "seeing" in a perception is primarily a vision of outlines, and integrates this with notions of feedback and feedforward. We literally perceive the forests before the trees. Not only do men like the poet William Blake see "outlines," as Bateson discussed in his metalogue, but we all do exactly that. Our initial percept, which can be called that of "vision at a glance," is one that constructs a generalized categorical scene interpretation. It has been preceded by a very rapid feedforward stemming from the lower levels of perceptual neurons, or from the bottom up, of the neuronal levels to perceptual hierarchy (since the cortical levels of perception are hierarchical) to an upper-level cortical "vision at a glance." It is followed by reverse hierarchy routines, which then focus attention on detailed information that is available at the lower neuronal levels. The initial feedforward effect is always implicit, or subconscious, and is automatic. The reverse hierarchy is a feedback down to these lower cortical levels, following perception at the top, and permits "vision with scrutiny," which overcomes initial "blindness" to details.

The theories of the Gestalt school on which Bateson had relied for much of his understanding of perception had emphasized immediate perception of global attributes through processes in which individual elements combined to form a whole—all accomplished without conscious awareness. RHT takes this a step further. It presents experimental studies to show that in explicit—that is, conscious—perception, the whole is perceived first. Our explicit (conscious) perception perceives words before letters. "Explicit perception" includes stimulus-driven experiences that are accessible to conscious identification and recognition, and/or retrieval from memory. Initial feedforward proceeds through a feedforward loop from the bottom-up hierarchical pathway, but not all of this in the feedforward process is available to conscious perception.

Instead, explicit visual perception only begins at higher cortical levels. It then proceeds in top-down fashion to encompass detailed information that is available. Details incorporated at these later stages of explicit (conscious) vision include precise location, retinal size and color and component motion (Hochstein and Ahissar 2002, 796). Initial explicit perception proceeds by taking an approximate "guess" as to the binding features falling within the same large high-level receptive field, but such features may arrive from separate streams of information. Given this initial crude binding, reentry through feedback confirms or refutes this initial "guess."

The general argument presented in RHT assumes, therefore, that explicit vision is achieved through congruent formation of categories, primarily through a sort of "guessing" of category from detailed information received implicitly. As we shall see in later chapters, Bateson borrowed terminology from C. S. Peirce, including the term "abduction" to account for a logic of "guessing" (Chapter 8). RHT holds that detailed information could include subordinate categories, which then require further identification through the reverse hierarchy, top-down phase, in moving toward "vision with scrutiny." Vision with scrutiny then proceeds to unbind initial incorrect conjunctions and revise "vision at a glance" when unexpected conjunctions are present.

Hochstein and Ahissar suggest that the principles of RHT might apply to perceptual learning as well as to perceptual process. Whether they are correct or not, it is evident that the universality of gestalt covers conditions other than that of the well-known universal propensity to distinguish figure and ground. The gestalt effect exhibits form—a property of perception whereby simple geometrical objects are recognized independently of their rotation, translation and scale, as well as elastic deformation, such as that which occurs in topology. Another gestalt feature is that of reification, in which the percept experience is deemed to contain more explicit spatial information than the sensory stimulus on which it is based. Gestalt reification usually refers to basic shapes, like the perception of triangles when no triangle is present. Herein lies a perceptual origin of the concept of the "fallacy of misplaced concreteness," the fallacy that Bateson had identified as afflicting anthropological researchers when they misread their own anthropological synthesis of cultural structure as being a "real" example of the way that culture persists. This example may also demonstrate another feature of gestalt, namely the "law of closure," in which the mind may experience elements it does not perceive directly through sensation in order to complete or increase regularity of a figure.

Bateson would refer to the "punctuation" of perceptual events. Punctuation introduces some kind of space, or time or topological boundary into a plenum, or continuum. Above all, gestalt proposes that the operational principle of neurological perception is holistic, and analogical, with self-organizing tendencies and parallel distribution. It also holds to the principle that the whole is greater than the sum of its parts, or, at least, that perception requires investigation of part and whole and their interrelationship. Later, Bateson's discussion of punctuation of percepts would show that a characteristic of the whole of life is its ability to learn through the contingencies of analogy in punctuated perceptions, though the characteristics of these analogical contingencies in relation to behavior may be different among any species of living organisms.

PATTERN IN INTERPERSONAL COMMUNICATION

We are now able, by way of summary, to link all these terms and phrases, showing their relationship to each other. Overall, Bateson thought that dynamic patterns of perception, rather than direct sensory input, generate patterns of the experienced present. At another time, the pattern experienced is anticipated once more, and at this moment (a second time around, or recursively) action is based on that expectation of a future state. In between lies the meaning of difference related to expectation of a future state. In his later years, Bateson held firmly to the view that difference is primary in perception. His position was to understand difference as "change" in the ever-shifting ground of informational flux, and not to take difference as a commonly framed antonym to resemblance, identity or similarity, which would, of course, push the terms "difference" and "similarity" toward a type of terminological dualism.

Another way of putting this is to say that perception enables distinction to be made between pattern and noise. The primary means for discrimination is that of perceptual comparison in the context of comparing changes; thus, pattern is associated with learning about contexts and learning with imagination. In any event, difference is something other than that which is to be found in the division or multiplication of identities or resemblances, once order in pattern is uncovered (MC, ca. 1976, 955–36a). While recognition of similarity and resemblances occurs with the ordering of pattern, the latter (pattern) has to be distinguished from noise before the ordering of pattern reveals identities and resemblances. "Our knowledge is never a knowledge of continuities but always a knowledge of heterogeneities," Bateson wrote in "What Every Schoolboy Knows," an unpublished manuscript. "Homogeneity [sameness] we know nothing about except by extrapolation from perceived difference."

Bateson's Terminology for Pattern in Interpersonal Communication

Pattern	"Now I suspect that when we talk about 'patterns,' we are not referring to the categories or sub-sets in such a classification, but rather are, or should be, referring to some *derivative* of this system of sets; and further, I have a hope that if we could define these derivatives, we should thereby define (and succinctly) some of our most elusive variables" (Letters, LOC, 11/04/1947). A derivative of a set of relationships among constituents is a pattern, and pattern identification is basic to any inquiry. Rules of various types are integral to the formation of pattern and its variance.
Relationship	Relationships are primary ordering principles. Until relationships are categorized, there is nothing but unorganized distinction. Types of relationship tell us how a message exchange is to be interpreted. A relationship is a connective principle. It is a rule that instructs, clarifies and explains how two things stand toward one another. Talk is organized by these connective rules, which also operate as performance rules.
Rules	Unlike most meaning or semiotic assumptions of communication as being the product of an *individual*'s intentional state, the notion of rules-as-triggers in interactive communication directs attention to rule governed interaction, which is the context an individual uses to create his subjective state in the first place.
Context	Context is form, not substance, and all meaning of message exchange should lie in reference to the formal properties of one exchange in relation to other message exchanges. Although context may be conceived in a ladder of context, metacontext, meta-metacontext, and so forth, the ladder of contextual nesting is best conceived as an intransitive part-whole relationship. Levels of context have binding properties, but because a context is an abstract form, its binding properties result from events that occur between communicants, and not necessarily from events that occur within individuals' minds or from logic that binds argument to its premises, as in rhetoric and mathematics.
Time	Interactive behavior plays an important role in the experience of time. Equally, time can be equated with communicative behavior. Time is experienced as a part of the formal properties of communicative events, and the experience of time can recur as communicative behavior recurs. So, a relationship is experienced according to the organization of the exchanges of the communicants and this relational quality is linked to both cyclic and idiosyncratic phases in time.

Difference	It is impossible to understand some phenomenal unit, a relationship or pattern of relationships, without referring to differences between that phenomenon and something else. Difference is perceptual, but it is also the basis of all comparison. Understanding differences is the key to making sense of that phenomenon, but making sense of difference is always done in relation to others. Difference and comparison of difference is primarily perceptual and formal. The world of "ideas" in its most elementary sense is synonymous with "difference" (Bateson 2000, 472). Percepts are available for cognitive consideration, with the "outside" perception linked to "inside cognition."

(Adapted from Ellis, in Wilder and Weakland 1981, 215–230)

In a later formulation in the unpublished manuscript "The Evolutionary Idea," Bateson said that only news of a difference can enter into man's sense organ, his mapping, and into his mind. Only difference can effect and trigger an end organ—so all information (our universe of perception) is built on differences:

> Difference is out there and precipitates coded or corresponding difference in the organism's mind. And that mind is imminent in matter which is partly inside the body—but also partly "outside" e.g. in the form of records, traces, etc. Difference is "super natural" i.e. outside the natural world as this is seen by the hard sciences. Yet difference is not located in x or y or in space between as Kant had proclaimed in his *Critique of Judgement*.

Thus, he is able to make a basic epistemological statement about the relationship between all reality out there and all perception here in the head.[1]

CONTEXTS AND HETERARCHY

One of Bateson's major themes, along with cybernetics, was that of holism. In 1957, he had written that if the holism of the eighteenth century had been the "great chain of being," a fixed and rationalistic hierarchy covering all ideas, the hierarchies of the nineteenth century had dovetailed materialism as expressed in biology, botany, geography and geology within the ladder of this great chain of being. The holism of the late twentieth century needed to change these perspectives. The new holism should rest upon the properties of information and levels of information. Its "great chain" would be dynamic, indeterminate, and not a fixed series of steps. It would express its levels or steps in terms of indeterminate processes of "becoming" rather than a fixed chain of "being."

And cybernetics would influence the shape or form of the new ladder because cybernetics would introduce a new system of systemic thought characterized by levels of communication that were circularly causal and subject to feedback.

Deriving any holistic perspective obviously requires an intensive investigation of theories of perception. Perceiving a laddering effect in the circularly causal levels of communication would therefore involve very different skills than an imaginary climbing the eighteenth century ladder of the great chain of being in which human step by step knowledge moved from bottom to top. As this chapter has shown, so far as Bateson's ladder of perception was concerned, the patterns of perception always involved a circularity of topness to bottomness and its reverse. Moreover, perception required envisioning between levels. The very act of perceiving required an ongoing filling-in of the details of perception an infill of topness into the loops of bottomness as movement or change occurred.

Finding analogies for presenting such movement in levels of perception was by no means an easy task. An early example of this had been Bateson's wrestle with Radcliffe-Brown over the latter's notion of social structure. It recurred again when Margaret Mead's organization of their Bali material took their joint research into the realm of culture and personality. Bateson was to reject this format after the war because he realized that the mistake they had both made was to fail to take into account the number of levels that intervene between percepts of individual personality and percepts created by a group dynamic, and which is expressed either consciously or nonconsciously through habit in cultural performance. As cultural performance changed after World War II, so did the complexity of the interrelation of culture with personality. An example might be of the fear of upside down gods. Traditionally in Bali, the spiritually dangerous areas where one might encounter an upside-down god are in graveyards or in forests where the monkeys dwell. Modern tourists would have no such fear and would welcome the provision of showers and washrooms built especially for them in these areas. The Balinese entrepreneur who transgressed cultural behavior in order to provide such facilities was so successful that he was soon able to open a restaurant, but he still honored the spirits that dwelled in the hearth of his house.

When Bateson comes around to a discussion of steps, or hierarchical order of levels, as he does in his papers on learning, the levels or steps are expressive of processes induced through changed contexts and/or percepts, and not of steps moving onward and upward in a supposed logical sequence. Bateson refers to a hierarchy of contexts that lack any physical location, or inertia, or momentum or any precise temporality—all of which would be measured in a physical system. Contexts are real to the extent that they are an effective part of the communication of messages, although they are not real in any physical sense.

The notion of context was crucial to any process of learning above that of mere rote learning. The relation of Learning I to Learning II is via a perception of difference in patterning—and the perception of difference then becomes self-validating (Bateson 2000, 289–294). The learning of Bateson's Learning II is a new way of punctuating events—a way of seeing inkblots in a different pattern, initially punctuated by a variety of stimuli and reinforcement. Another of Bateson's references to patterning and learning is as follows:

> I had some vague idea that to set up "sameness" as the other half of "difference," would promote all the fallacies of misplaced concreteness which all nouns encourage. The point is that emphasis on difference leads to *patterns*, emphasis on sameness leads to *quantification*. [In the latter case] There is always the question "how much sameness?" (MC, 12/05/1970, 1519–8f)

Metacontext Recursion

Bateson believed that his ladder of informational contexts and metacontexts were a great improvement over the notions of traces or other theories of physical embodiment had previously dominated Western thinking. His levels in the ladder of contexts are derived from interplay between messages at different levels of experience, including play, art, and ritual or any means by which messages are interpreted in context. The derivation of context does not have to derive from speech, but if it does, it is meta to how the words are phrased; context is connotation as distinct from denotation, with the relative invariance of contexts-as-connotation, compared to the rapid phrasing of the words in the communication exchange. A ladder of contexts is that of connotation of message, metamessage, and meta-metamessage.

Thus his hierarchical order is never truly "hierarchical" in the manner of physical systems of control like bureaucracy or paramilitary command posts. The dynamics of context recognition emerge from systems in which higher and lower levels of dynamic circularities may in some way constrain each another, but are not determined by one another, as in physical systems. Rather, there is interplay between different levels of context and propositional order, and with different levels of learning derived through processes of feedback entering into the pattern of communication. Patterns of order in each level are also multivalued, so that "top" and "bottom" are relational-outcomes of specific relations, rather than being thought of as a coordinated command structure with little variance.

Bateson's illustrative representation of context meeting context indicating context, then metacontext, then meta-metacontext, and so forth yields a topology of multiple cybernetic circuits; or in its ecological version, as parts of a whole in another relation to other parts of the whole, which always includes the whole ecosystem in its systematic relations of parts in a whole. What then are the ways in which context is also related to gestalt and thus to processes of perception? Perhaps "in a context" is not the best means of expressing a meta; rather, events are considered to be parts of contexts that they help compose, contributing to a recursive series. His ladder of context derived from constant repetitions of external communicative interaction, or internal memory of such communication, or both, thus transforming what engineers call "redundancy" into what Bateson called "pattern" as the primary phenomenon. As the quote at the head of this book indicates, informational redundancy reveals a pattern that is usually recognized at a point where it meets another pattern at which point it creates a third pattern of contrast between the two, which then achieve a broader or more holistic dimension of understanding about the characteristics of both patterns and a wider gestalt embracing the contrast between them. It is not a ladder in which primal conditions predicate the next step. A problem therefore existed in Bateson's using the term "hierarchy" to explain his notion of levels when he did not mean "hierarchy" in the accepted sense. It forced him to explain in *Steps to an Ecology of Mind* that his ladder was *not* built on the transitive hierarchies of logic and mathematics where each step or level necessarily followed the initial premises or hypothesis or the bottom. This is not what he meant at all. At the same time, as we shall see in the next chapter, he was using Bertrand Russell's notion of set theory and logical types, which was indeed built on the transitive hierarchy of class and its members. The problem arose because Bateson derived his understanding of "hierarchy" from McCulloch's explanation of the relations of redundancy—a technical term in information theory—contained in the circular circuits of short-term memory and long-term memory.

Redundancy, as McCulloch presented it, is polydimensional, and therefore McCulloch chose the term "heterarchy" rather than "hierarchy" to describe the polydimensional nature of the patterning. McCulloch's term "heterarchy" suggests a play of ideas occurring in and between two circuits of information, one of them being long-term memory and the other short-term memory. Bateson took McCulloch's research to indicate how human beings learn, and how we can learn to learn about the multiple patterns in which we are involved. If the mental characteristics of a living system are immanent and systematic—not in some part, but in the system circuits as a whole—then educators are misrepresenting the degree to which the individuals' insights are autonomous, a radical critique in the educational epistemology of the West.

Essentially, a heterarchical conception of the formation of choice is Bateson's categorical reply to the soldiers of reason discussed in the previous chapter— meaning that the "soldiers of reason" mode of evaluating choice through strictly defined criteria of utilities or rationality is thoroughly reductionist. Yet the polydimensional aspects of context require other images to aid its comprehension. On occasion, Bateson suggested a Wittgenstein image of the existence of "family resemblances," defined as overall pattern that runs through the whole thread, or the continuous overlapping of those fibers (Harries-Jones 1995, 177). The metaphor implies that meaning could not emanate from a single, reverberating transmission of a defined message. Information as news in a circuit responds to *all circuits oscillating in a system of overlapping fibers*. Our understanding of contexts should relate to the way that we embody thinking and acting in external interactive settings together with any known internal loops, and all of these, in turn, are coupled to our assumptions, and to how we learn. This is true holism.

Typically, McCulloch's notion of heterarchical interaction in the relation of short-term memory and long-term memory suggests complementarities of relationship occurring at each level. His notion of a "heterarchy" of contexts meant a brief temporal dominance of a series of contextual correlations that helped make a verbal message meaningful, or established a meaningful pattern through any other sense or combination of senses. A heterarchy of nesting contexts provides a means for self-correction both up and down the nest of levels in communication discourse made possible because the nesting context is mutually engaging and supportive. Altering the regress of nested contexts in context attribution can also create a path for communicational correction (Bateson 2000, 279–308); thus the difference in a heterarchy achieving stability as compared to a hierarchy. In so doing, self-correction avoids both extremes, namely downward forms of reductionism, where holism is seen as providing no new information, and upward reductionism where abstract forms of ideological holism replace the variance and indeterminism of meaning, so that interactive parts are seen as providing no new information (Bredo 1989, 29)

If pattern formation is as polydimensional as described, then, as subsequent mathematicians have realized, no actual graph can be drawn, and if no actual graph can be drawn, then it is impossible to measure and quantify choice in such finely observed polydimensions. To measure and quantify choice, transitivity rules have to be identified and held in a strict logical manner; if they do not, then there can be no graph, and no network of choice derived from such a graph. A mathematical solution to polydimensional choice under these circumstances would require an n-dimensional model, with a complexity of infinite centers and peripheries, all remaining linked with all the others, and in network alignment, with continually changing relationships. There cannot be any single quantitative measurement of this complexity in these circumstances.

The mathematicians Eberhard von Goldhammer and Joachim Paul (2007, 1000) argue that the scientific mainstream in artificial intelligence has not sufficiently taken into account either McCulloch's initial study or Bateson's study of "logical" categories of learning, despite the fact that both are basic studies more than fifty years old, which "from a methodological point of view—are still unexcelled" (1007). And at least so far as the existence of pattern in communication is concerned, heterarchy of contexts can only be described qualitatively, as differences of contextual situations. We will return to the issue of mutual complementarity in the heterarchy of ecological contexts in Chapter 7.

5

THE BONDS THAT BIND

As we saw in the previous chapter, Bateson proposed that communicative contexts appear in "a ladder," or "a hierarchy," or a "heterarchy," but that defining what sort of ladder this might represent is a problem. Context is "meta" to the content of the message, but unlike the steps of logical or mathematical inference in which strict procedures apply at each level, individuals responding to each other in communication contexts may punctuate levels of serial communications differently.

Bateson's understanding of communication was that it is an event that contains a second-order pattern, that of context, which frames the relationship between the communicators. The second-order pattern gives meaning "meta" to the contents of what is actually said in the message exchange. Second-order patterns are especially prominent in family process, where relationships order family life and communication exchange is highly repetitive and highly redundant. Communication within the family is a connective principle stemming outward from the immediate speakers to a network of involved others. If the hierarchies involved were strict logical or mathematical hierarchies, they would, by definition, be transitive hierarchies—that is to say, the rules of relation would be strictly ordered. Bateson's schema of contextual levels are drawn into intransitive heterarchies, in which the pattern of redundancies creating any one context can be recalled from events occurring all over the place moreover, these events are not selected, or fitted together in any particular order. For these reasons, attempts to correct communication in social situations is difficult for human beings to undertake by themselves, since it involves several layers of relationship of self-with-others all at the same time. The variety of intransitive patterns in the tapestry of family organization is extremely difficult to perceive, and even more difficult to perceive when they have become a matter of habit.

Yet all communicators are bound into habituated interactions, derived from the experience of past communications and the expectation that current communication will repeat itself. The percepts that form a gestalt can include feedback response as well. We duck if we see an object flying toward us, or we habitually respond with an apology if we inadvertently bump or nearly bump into another person. An injunctive rule of "don't ever apologize" in near-bumping incidents would be difficult to sustain. The unconsciousness or nonconsciousness of habituation makes refocusing of communicative interaction a difficult task. A person can get very easily into a tangle or a "bind" in his or her communication with others, a bind that began as a "horizontal" bind of personal interaction in individual family circumstances, but can later can develop a "vertical" systemic bind. A horizontal bind could be of the sort in which the content of a message about "I love you" is continually refuted by the context in which the message is communicated. That is worrying in and of itself. Yet vertical binds frequently occur in which participants find that it is not possible to talk about the "bind" that they face, and where having a "resolute up-front" conversation is unlikely even forbidden. Continual repetition of the vertical bind continues, yet family relations are such that often there is little possibility of being able to get out of the relationship between binder and victim: once a son or a daughter, always a son or daughter.

For any researcher into the family system, uncovering the rules of relationship that give context to repetitive communication events is crucial because "the astonishing repetitiveness of mammalian discourse about patterns of relationship follows . . . the need for definiteness which is endlessly reactivated by any silence on the subject matter of relationship. So it is never enough for human beings to agree that they love, hate, respect etc. each other. They must be forever re-affirming that fact" (MC, 18/05/1965, 1399–1c). It is characteristic of family members to share not only rules of interaction but also for the redundancies of interaction to develop gestalt about patterns of communication, a gestalt that includes contexts of communication. The context of a communicative interaction couples communicators to each other in two ways: first, in creating order and/or disorder in the family system; and second, through a memory of these same ordering events, which enter into other contexts. In this respect, communication events are quite unlike energy events. Through these patterns in ongoing communication, any communicative situation implicates prior feedback and expectancy for future communication.

In other words, the cluster of repetitive or redundant contexts of communication with regard to relationships does not create a coding of information of messages transmitted from sender to receiver as per the Shannon-Weaver definition. Instead, the formation of a gestalt is a derivative of percepts about style of communications common to relatives in the family, and also an understanding of feedback to response among family members. Another way of putting the same idea is that contexts of communication create the communicator's subjective state for responding in the first place. Recall Bateson's definition of pattern provided in the table in the previous chapter: "Now I suspect that when we talk about 'patterns,' we are not referring to the categories or sub-sets in such a classification, but rather are, or should be, referring to some *derivative* of this system of sets; and further, I have a hope that if we could define these derivatives, we should thereby define (and succinctly) some of our most elusive variables" (Letters, LOC, 11/04/1947).

SCHIZOPHRENIA

By far the most notable contribution of the Bateson Research Group at the Veterans Administration hospital in Palo Alto was the study of the way in which the malady of schizophrenia affected communication activity among and between family members—and to the pathology of what Bateson and his team called "double bind." Bateson had some familiarity with research on schizophrenia before coming to Palo Alto. Margaret Mead had received some research money to study aspects of schizophrenia in Bali, and though it would appear that this never became a high priority in their research the duo had contacted Dutch specialists and enlisted the help of Jane Belo in preliminary research. The duo planned together "a great interdisciplinary expedition, complete with endocrinologists and psychiatrists, which was to come to Bali and have a headquarters in Bangli Palace" in order to undertake a study of schizophrenia. "We even took a three-year lease on the palace, although we knew that the plan represented no more than a dream," because war in Europe seemed imminent (Mead 1947, 259).

It was indeed fortunate that Bateson and Mead were not able to sponsor such a project. In those days the psychiatrists believed that schizophrenia was a return to a "primitive" form of mental functioning in which the effects of civilization had not yet overlaid the primary processes of mental life through rationality. Instead, schizophrenics subject themselves to a regressive state of unfettered emotionalism and lack of self-awareness. Yet the evidence Mead and Bateson had already gathered from Bali would not support such a view. It was a lack of balance that the Balinese feared. Balinese people encountering an upside-down god expressed their fear of having dreams in which they saw their bodies flying into fragmented parts, dreams that implicated their own loss of integrity.

Modern psychiatry no long supports these older ideas, though in Bateson's time psychiatrists were still speaking of schizophrenia in terms of "loss of emotional discrimination." Today modern psychiatry supports the view of biochemical imbalance in the brain as a possible cause (the dopamine hypothesis) plus the idea that the experience of schizophrenics commonly reveals misplaced modes of abstraction. One recent study argues that the abstractive pattern of schizophrenics is quite the reverse of regressive primitive emotionalism and is much like that of a mindset of a hyperrationalist. When schizophrenics reveal their experience of their own body, they are devoid of bodily senses and imagination, and like unrepentant followers of Descartes, they talk of living things in general as mechanisms, or simulacra, or assemblages of independently moving fragments. Schizophrenic subjects see themselves as robots, computers, or cameras and believe that parts of themselves have become replaced by metal or electronic components (McGilchrist 2009, 192, 351, 398). They see the world around them as vaguely threatening.

Bateson was reintroduced to the issues surrounding schizophrenia after taking up residence in 1949 as an ethnologist at the Veterans Administration hospital, and he was to stay on undertaking research until 1963. He began by undertaking an ethnography of the social organization in the hospital, which narrowed to studying therapeutic practices at the hospital and, later, to being a therapist himself. When he had gathered research grants and a research team, he began to consider therapeutic practices specifically related to schizophrenia. This research began "all over the place," as one member of his research team, the young anthropologist John Weakland, put it. Bateson went back to the fundamentals of signaling in cybernetics to reformulate their boundaries of relevance for this task. Then, as we have seen, he drew ideas from game theory, gestalt psychology, theories of small group interactionism, and theories of learning in order to make his case (Harries-Jones 1995, 127–129). Weakland described their research as a process of digging themselves out of holes previously dug by received wisdom that had prevented people from getting anywhere with an understanding of schizophrenia (Wilder and Weakland 1981, 46).

Bateson's research team noted that the behavior of schizophrenics was both fixed and mechanistic, but took the overall view that schizophrenic behavior was to a large extent associated with the inability to perceive appropriate context in communication. The schizophrenic subject was unable to discriminate appropriate signs and signals, and it would seem that such inability was in part learned from communicative patterns in the family situation. Therefore, patterns of communication in the family setting might be an important part of any therapeutic reframing.

Typically, schizophrenics will eliminate from their messages everything that refers explicitly or implicitly to the relationship between themselves and the people they are addressing; therefore their means for understanding and classifying contexts of behavior are considerably diminished.

Bateson had noted that conditions of schizophrenia arise out of group dynamics in which schizophrenics find that they confront situations in which their part in patterns of communicative coupling, that is the premises and the contexts upon which they assume "relations between," are constantly violated. This results in a situation in which "he (the schizophrenic) may eliminate anything which refers to, and is part of, a relationship between two identifiable people. . . . He may avoid anything which might enable the other to interpret what he says. He may obscure the fact that he is speaking in metaphor or in some special code, and he is likely to distort all reference to time and place" (2000, 235). The schizophrenic will eliminate everything that refers explicitly to the relationship between the person addressed and himself or herself. This pattern of response suggests that schizophrenics do not overtly communicate because they believe that they will be punished every time that they are right, in their view, regarding the context of their own message.

Psychiatric terminology called this aspect of schizophrenia "breakdown in discrimination," but Bateson proposed a different sort of process, one in which the afflicted persons have found their contexts of learning falsified or disabused so that discontinuities arise along levels, or boundaries, where they have suffered disabused expectancy, trust, reciprocation, and love. This, in turn, often results in violence toward the self. As the pathology increases, an increasing gap emerges between the victim's perception and the victim's conception of a state of affairs. In a full-blown psychotic state, an individual may be unable to tell the differences between evident differences. In other words, the differences between something being X and not being X has ceased to matter. The same with communication: the desire to communicate about experience and yet feeling frozen in any communication about experience is, they feel, one and the same. The victims may be aware that they can navigate space and move in time, but none of it feels like it is happening to them, or that it makes any difference to their present situation.

Once in a pathological state it is very difficult for the victim to find a way out, for it can only be brought to an end by the very uncertain means of the victim regaining "insight." Recognizing patterns of feedback is in itself a symptom of remission in pathological condition. Bateson pointed this out in his introduction to a book tracing the onset, treatment and remission of a schizophrenic, who in the nineteenth century wrote an autobiographical account of his condition and his eventual cure. Perceval, the schizophrenic, was aware enough during the course of his episodes to keep notes on his response to the doctoring he was given. Bateson's commentary *Perceval's Narrative* (Bateson

1961) showed that "cure" did not rest specifically with the doctoring received, but rather with the patient slowly coming to understand how he himself can perceive onset of malady.

In his commentary on Perceval's narrative, he points out that the primary task of mental care should be aiding a self-cure; caring for the patient should continue until the patient is able to reorganize his or her mental states. The experience of such "a curative nightmare" is always visceral, a "curative nightmare," that seems to recapitulate the dynamics of death and rebirth, since body or mind, or both, seem to attack the psychological self. For some who survive the "curative nightmare" the near-death experience seems to correct a deeper pathology. The correction of the deeper pathology, in turn, enables a move toward remission or other minor forms of resolution (Bateson 1961, xii).

LOGICAL TYPING: A LOGIC OF RELATIONSHIP

To refocus or reframe in a therapeutic context is to reorganize a cluster of differences arising from habituated patterns of communicative interaction and to consciously form a new pattern of interpretation. The communicator perceives that differences in perception make a difference in communication, also that the communicator can alter his or her interpretation of events and that the process of response to response creates a classification of types about such clusters of differences. In more formal terms, there are links between the proposition, perception, learning, and gestalt, but propositional links about difference or movement or change of some sort must also include a perception of those differences. The presence of discontinuities allows questions and answers about patterning, not only in the sense of inclusion and exclusion but also in the sense of how different aspects of "logical" types interpenetrate—how they include or type each other.

"Logical types" is one of those slippery terms that Bateson sometimes uses, that are slippery because they are analogies rather than classifications. As a concept of classification, logical typing is a form of distinction indicating a use of a different set of criteria, or classes of phenomena, when an observer undertakes research on a complex issue. But Bateson introduced the notion of logical types as indicating a form of distinction among subjective propositions. By this means, logical types and the differences between them become related to propositions of belief about relationship, rather than indicating objective types of classification, neither "objective" in its original sense nor subjective whims in patterns of relationship. It was a logic of distinction in between subjectivity and objectivity, subjectivity in the form of reports by clients (patients) in therapy and "objective" in the form of a class of events that can be utilized by the therapist. Then he associated the use of this term with variety of levels in communication contexts with other persons in the family setting, so that logical types create levels of "punctuations" in perception: "As I use 'higher logical type' (HLT) the relation is commonly between propositions [of belief] rather than between items, classes, names, etc. Propositions as members of classes of propositions—and there is a sticky, slippery edge. Can a class of propositions be itself a proposition? . . . I guess I sinned in pushing HLT away from names and classes *without saying* I was doing this" (MC, 14/06/1979, 591.5–2a).

A logical type poses the question of how any particular pattern at any particular level is perceived. Logical typing is also an ingredient in the relationship between any description (that of the therapist included) and any system to be described. Subjectively, it is linked both to perception and to learning and to the notion of differences that make a difference as summarized in the following Bateson comment made a decade after leaving Palo Alto: "In logical typing we are concerned with differences between differences or differences in how the difference is going to belong. . . . It is a learning response [in triggering an 'a-ha'] to a larger gestalten than you had previously used. That's it. You suddenly see the larger pattern" (Dialogue between G. B. and Werner Erhard, 09/03/1976 and 09/06/1976).

As noted, the "logic" of these contextual levels does not follow the fixed procedures of either rhetoric or mathematical proofs, but therapists and their patients can derive them from an understanding of interplay between messages at different levels of experience, including play, art and ritual, or any other means by which a message can be interpreted as belonging to a particular context, which they can then classify. Each logical type represents a relationship. The relation of logical typing to therapy comes about because therapeutic interventions are considered to be parts of contexts of communication, which the therapy helps to compose. The therapy contributes to a recursive series of communicative events between therapist and client that can then be analyzed according to a logical type.

Originally, Bateson drew upon a mathematical approach devised by the philosopher, Bertrand Russell, and used Russell's approach to give gravity and validity to his scheme of logical types. Russell held that whenever we are engaged in rigorous argument, we must only include within a given class those items that lie within that class of the same level of abstraction, and never include an item of another logical type, or abstraction. His example was that of the distinction which lies between a singular person and of a class of people. There is a class, and there is a classifying line between the class and its individual members, as found in the most common statement "but of course I am not including you in what I have just said about class Xx."

Any type in the abstract notions of the mathematics of set theory is very brittle, such that if it is breached it leads directly to a paradoxical mathematical assertion. This is because a type is regarded as either A (within the bounds of the set) or B (outside the bounds of a set). A popular version of a logical paradox derived from Russell's conundrum is that of the barber who shaves all the men in the village who do not shave themselves—in which case who shaves the barber? This led Bateson to an analogy between the conditions of paradox in a mathematical or logical statement and the sort of paradoxical situation that a pathological family always confronts.

The problem is that while Russell's rule can be enunciated in mathematics as a strict rule of logical order, the nonlinear and circular world of communication with many levels, where communicative events return to their point of origin, creates situations where strict reference to a class and its members become displaced. In this circumstance, it is virtually impossible to avoid some types of paradoxical communication, in fact normal communicative behavior breaks Russell's rule all the time. Bateson himself had to admit that Russell's logic does not apply literally and that, in pragmatic circumstances, conversation consistently violates class/member typing (2000, 189, 203).

When communicators find themselves in a quandary in their communicative relations with other people through breaking Russell's rule, they learn, under normal circumstances, to manage what do with broken rules of communication. Many times a day people misuse psychological ego function—that is the confusion of "we" and "they" with ego, "I"—without ever being aware of it, unless social conventions or social prohibitions draw this to their attention.

Any message that qualifies another is of a Higher Logical Type (HLT). A simple illustration of "logical type" is the difference perceived between ego's verbal statement of a "truth" or a proposition, and ego's "body language" when making "truthful" statements. Yet the most simple examples of logical types (kinesic and linguistic) were the ones that the Bateson team of researchers found most difficult to draw any conclusions about, because it was never clear to them as to which logical type was "meta" to the other. As Haley noted out in his long paper on the history of the research: "kinesic and linguistic patterns proved too difficult to bring within the theoretical framework, perhaps because they were devised for the study of the individual rather than patterns of relationship [in the structure of family interaction]" (MC, 09/7/1970, 593–103b).

Yet second-order levels, ones revealing the logical type of patterns of communicative interaction, had much greater salience. Here the analogy that Bateson drew between psychological process and mathematical procedures is meaningful to a therapist's understanding of critical relationships in the system of family interactions. Therapists could set about "mapping" perception patterns within the family and reporting them to the family in order to create awareness of how vicious circles are initiated. They can observe family rules about transforms in perception from class to member and member to class, as well as refusal to transform the perturbing patterns setting vicious circles into motion.

Drawing from the analogy to Bertrand Russell's rules about logical argument, therapists could suggest how a change in sequence and patterns of communication could avoid destructive patterns and "frozen" habits of communication. At a simple level, the therapist might suggest alterations in communication sequences among family members, or they could suggest that the family pay attention to the "timing" of communication among family members, or learn how to avoid habituated patterns of confrontation among family members. At a more complex level, there are several other ways of disrupting vicious cycles of interaction, including deliberately creating a "paradoxical intervention." Pathology can arise in a family system when sheer experience of continual patterns of paradoxical-type communication lead to vicious circles of interaction. The solution the therapist hopes for in "prescribing a paradox," is that the family becomes aware of how these destructive interactions arise in the routines of daily living, and daring the family to continue its self-destruction. It also can reveal how difficult it is for any one single family member to attempt to control these interactions from arising in the first place.

DOUBLE BIND

Bateson and his team associated communicative binds with schizophrenic behavior, but they insisted that there was no strict causal connection, though critics of the research team in Palo Alto alleged that they were indeed making claims about causality in their research on schizophrenia. In his "Four Lectures" (MC, CAF 126), as yet unpublished, Bateson discusses his ideas about logical typing and how it enters into a discussion of schizophrenia. Characteristically, "schizophrenic" families display extraordinary inconsistencies in their communication with each other, which as a family unit they regard as consistent and logical. The researchers asked themselves: What type of sequences of interpersonal relationships would induce such unconventional habits of response to each other? Their answer was a set of circumstances in which one member of a family has come to believe that they are in an unresolvable bind, because all choices they make in responding to interpersonal relations are untenable. Long lasting patterns of communication always create habitual expectations, especially if conditions within a social unit arise in which a member or members feel bankrupt in making a choice. Blocking choice then gives rise to seeming paralysis in undertaking communication, a "paralysis" commonly demonstrated among members of a schizophrenic family. The "schizophrenic" member shows almost complete inertia with respect to social relations and is noncommunicative, yet contextually noncommunication between the schizophrenic and other family members will, in some sense, be appropriate.

The research team acknowledged that single binds, or more straightforward dilemmas, could be either painful or traumatic, or could be just plain funny, but a pattern of a double bind, by contrast, is always painful and often destructive. While double binds can occur in normal, as much as in pathological, situations, the research team noted a strong association between the noncommunication of schizophrenics and the distinctive interactions of the family environment, which accounted for schizophrenics' feeling as if they are on the spot at all times. Problems were seen to arise from an inability to deal with the initiating conditions, and inabilities in learning how to learn, or deutero-learning.

The notion of "double bind" was by far the most notable contribution of the Bateson Research Group. Apart from the originality of its approach to pathology of communication double bind gave the English language a new phrase for use in general conversation. The colloquial interpretation of double bind has come to mean something like "damned if you do, damned if you don't." But colloquial interpretation depicts the circumstances of a single bind, rather than a double bind. The original expression of double bind refers to a much more complex set of circumstances. If faced with a "damned if he does and damned if he doesn't" situation, a person may choose the lesser of two evils. But when faced with two conflicting levels of message that are in some way injunctive, so that she cannot obey one without the other—as in the conundrum "I want you to disobey me" where if she obeys, she is disobeying, and if she disobeys, she is obeying—from a subjective view of the "victim," any move on the part of that "victim" disqualifies the interaction that the "victim" undertakes. Thus, the research team's attention began to concentrate on the injunctive aspects, of a process of continual going-around or oscillation from one pattern of feedback to another pattern of feedback that the former always evoked. The oscillating pattern is equivalent to a going around in circles from which the family could never seem to extract themselves, and required a change in the initial set of conditions, but this also seemed to be blocked.

Double bind concerns contexts of behavior between self and other, or self and environment, or even self and self at another time. Some of these contexts can be altered through experience, but in cases of double bind, any change in the context of existing communication patterns seems to threaten the very survival of the "self" in relation to all other relationships within the family.

At first, the research team gave only limited examples of double bind within a family, dyads of mother-son interaction patterns in which the "mother" twisted the meaning of "love" in relation to her son. The sequence might begin with an assertion by mother to son at one level of meaning, but at the same time also contain a negation at another level of meaning. Thus the mother says: "Do not do so and so, or I will punish you" (level one—message content); while, at the same time, saying: "Do not see this as punishment" or "Do not question my love of you when I tell you not to do so and so and I punish you" (second level interpersonal message negating the first); and, also at the same time: "Don't talk back to me" (child is unable to comment on the paradox in the messages being expressed). The child thus becomes caught up in rules about approved and disapproved behavior that are mutually negating; and, the child is either forbidden, or in no position to comment on the situation (Bateson 2000, 201–227, 271–278). And the pattern of the bind then affected more than mother and child.

A problem with explication of double bind was that the research team did not make sufficiently clear in their publications that "relationship" always constitutes the context. It is the relationship that is of a "different logical type" and not specific messages that may be found within the relationship. Initially they, and other subsequent commentators, treated the relationship and the message aspects of the double bind as separate components on the same logical level, thus giving the impression that the double contradiction arises is between messages rather than message and relationship. Even sympathetic reviewers like Watzlawick, Beavin, and Jackson (1967) and R. D. Laing (1967) were prone to this interpretive mistake (Wynne 1969, 52–70).

Double binds occur where there is no escape from the relationship and where conditions of "lock-in" provide no escape. They occur where there is communicative closure, as in family systems, or in therapeutic situations in hospitals, or cultural double binds, where there can be no stepping out from the bonds of relatedness between communicators. Relations of the communicators in a double bind situation are often a result of a dominant-submissive relation. This means that the "bind" includes not only "horizontal" patterns of communication and feedback, but also "vertical" ordered relationships of disparity among the other members of the communicative group. In addition, the closure of the system creates contexts and constraints that eliminate or neutralize possibilities of communicating about the double bind.

Blocked reflexivity, or inability to find another starting point, occurs because none of the parties dare attempt metacommunication, for the consequences for doing so would seem to be too destructive of their own selfhood in their pattern of interaction within systemic relationships within the family, i.e., as husband and wife, mother and child, father and child. Any communication on the victim's part to undertake metacommunication is disqualified, while other attempts to metacommunicate usually end up in becoming part of a sequence of destructive emotional feedback. In a more general sense, double bind can, therefore, be considered as an example in which the "victim" is repeatedly required to support both sides of a "paradox" in a relationship.

Double binding patterns of communication would not be so destructive were it not for their constant repetitions that affect a victim's understanding of communication contexts and hence modes of classification of context. Over time, participants in the double bind can become less and less capable of handling the situation, while the vicious circle of supporting both sides of the bind becomes more and more dysfunctional. If the family involved is dysfunctional in a communicative sense, dysfunctional feedback itself becomes learned, and the outcome can lead to a situation in which the various participants appear to be powerless to act on or restrain outcomes, and so the family will interact in a way that maintains the pathology of the bind. In addition, the family communication process seems to find the context of pathological communication logical, plausible and consistent, thus reinforcing the overall pattern of events. Only "the victim" finds the situation untenable or absurd.

TRANSCONTEXTUALITY

The orientation of the Bateson project was anthropological, and it avoided any experiments simulating double bind in typical laboratory conditions. The research group preferred to approach a problem by observing situations with as little intrusion on the data as possible. In keeping with this anthropological orientation, Bateson objected to confining double bind to specific interpersonal situations and preferred to use the term in case studies where its significance could be compared and elaborated to disparate phenomena: "From this point of view the double bind is applicable to evolutionary processes and telencephalization of the brain, while it was also a specific entity in family relations which induced severe trauma" (MC, 07/09/1970, 593–103b).

This perspective of double bind as a general phenomenon, rather than a theory that limited itself to the more specific conditions of therapy, was to bring about a split in the research team. Jay Haley, for one, opposed the idea that research resources become directed toward any extension of the relevance of double bind. The incipient split accounts for one of the more unusual essays published in *Steps to an Ecology of Mind*, "Minimal Requirements for a Theory of Schizophrenia" (2000, 244–270), where the essay takes a long digression through biology and evolutionary theory in an implied rebuttal of Haley's position. By implication, Bateson suggests that Haley's case for confining determinants of schizophrenia to behavioral components was to introduce an idea of a template, and that this was similar to presenting schizophrenia as being determined through genetics alone. Determining factors for conditions such as schizophrenia are unlikely to be singular. Genetics may indeed control sequences of events at the level of the genes, but it has only partial control over events that occur at other levels. More specifically, he argues that standard approaches in genetics assume too easily the notion that the genome is a template controlling the substantive circumstances of life through a gene-script.

In the case of schizophrenia, Bateson argues, the natural history of the disease indicates that there is no direct chain of causes and effects from gene-script to phenotype. Rather there is a difference in the relative levels of determinacy of genetic, and behavioral, and environmental contexts (environmental here referring to surround of family network, more on which follows). He goes on to argue for a many-level or many-value approach that leads to a difference of response to schizophrenia during the course of the pathology. "Environment" enters in through factors that "themselves are likely to be modified by the subject's behavior whenever behavior related to schizophrenia starts to appear" (2000, 259). Since it is extremely difficult to find a genetic base for any single behavioral characteristic, the problems of schizophrenia must be approached by way of a theory of levels or logical types, a transcontextual syndrome, and "in terms of a hierarchic system in which change occurs at the boundary points between the segments of the hierarchy" (2000, 264).

His argument about "minimal requirements" of schizophrenia leads him to a rare, but brief, note on cosmology, or as near as Bateson gets to cosmology, under the heading "What is man?" Here he contrasts his own approach of transcontextuality, or holism, to that of standard assumptions in genetics, which propose that genetic activity in phenotypes (i.e., individual organisms at any one evolutionary moment) arise from genetic contexts that are isolable from environment. Standard assumptions arise from materialist science, a material science that declares that genetics is isolable from social behavior. Against this notion of isolable context separating the individual from his or her environmental surround, Bateson argues that while this may be true of matter, it is not the case with information and ideas stemming from information. He puts forward the notion that there is an infinite regress of communicative contexts linked to each other in a complex network of metarelations throughout the living world.

Instead of an infinite regress of contexts at differing levels of interaction, he claims that segmented separate levels, such as phenotype and genotype, are connected in a larger metacontext: "I have let in the notion of a universe much more unified—and in that sense much more mystical—than the conventional universe of nonmoral materialism." Acknowledgment of a metacontext brings about a new means of contrast between part and whole, namely that the whole is always in a metarelationship to its parts. This understanding offers "new grounds for hope that that science might answer moral and aesthetic questions" (*Steps* [1972] 2000, 267).

From this holistic and polydimensional perspective, schizophrenia was related to learning, to inappropriate judgment in deutero- (contextual) learning, and to behavioral interactions around relationship within the family system that fostered this. A triad of conditions affected its outcome: "I want to deal with questions about the impact of an experiential theory of schizophrenia upon that triad of related sciences, learning theory, genetics, and evolution" (Bateson 2000, 245ff.). As he was to write later, the onset of double bind was a result of some form of tangle in the rules of making appropriate transforms in a "transcontextual syndrome": "The mind contains only transforms, percepts, images and rules for making these transforms, percepts, etc. Since the rules are certainly not commonly explicit in conscious 'thoughts' presumably the rules are embodied in the very machinery which creates the transforms. There must also be a genetic component in the etiology of this 'transcontextual syndrome'" (MC, 23/06/1969, 52–6). In effect, he was initiating a transdisciplinary approach that would embrace communication, social theory, evolutionary theory, genetics, and ecology. All of this was disciplinary boundary crossing on a grand scale. Yet, this early version of his "ecology of mind" could not shake off the deficiencies of Russell's theory of types. Whatever the benefits in using Russell's logic as an analogy, he could not avoid the fact that Russell's logical approach went against his own definitions of communication as a recursive, rather than as a lineal order:

> Russell's Theory of Types, then, was a gallant attempt to create the logical frame for a non-recursive universe, one in which the relation between explanatory concepts could always be a transitive relation between "higher" logical levels and lower. And Russell was wrong in the sense that the world in which we live is not organized in lineal, transitive relations. If the chains of causation are to be plotted on to something like the chains of [an organic] logic, then the model must be a recursive one. (Bateson, "The Evolutionary Idea"—Glossary Box 5 Loose Manuscripts)

He would turn away from Russell's representation of types toward other schema in order to portray discontinuities and recursiveness. Recursion in relations of parts and wholes, and how they interleaved with each other required a different sort of algebra than that of set theory. His evocation of logical types had been a short cut to portraying how differences, contrasts, continuities, and discontinuities might arise at boundaries or levels of learning. Any perceptual awareness of "difference" required a boundary, a contrast, or perhaps a rotation of the other in order to arise in the first instance, so that recognition of a discontinuity required at least two differentiated differences, wedded together (MC, 1971, CAF 308). Boundaries could not emerge from a singular phenomenon, so that any boundary always referred to a pair of differences among coupled communicators. We shall consider further the transitions in his thinking—about pathology in ecological levels of "transcontextuality," as he now called it—in Chapter 6.

PROBLEMS AT PALO ALTO

Bateson decided to leave the colleagues of his Palo Alto Research team in 1966. His departure was followed shortly thereafter by the death of Don Jackson, the other senior member of the team. The Brief Center of the Mental Research Institute in Palo Alto was established in order to advance Bateson's therapeutic techniques, and opened approximately at the time that Bateson decided to leave. By 1974, the initial research team and its newer members at the MRI had produced nine books and 130 articles on systemic family processes, yet despite this publishing success, members of the research team continued to drift away. In the wake of this drift, a number of critiques began to emerge about MRI's conceptual approach to therapy.

Part of the critique flowed from the sheer novelty of the Palo Alto approach to the study of communication. John Weakland, when asked in the mid-1960s to summarize the direction of the Palo Alto team's research, found that that writing about their research required justifying the necessity for any theoretical approach to communication in the first instance (Weakland 1967). It took about ten years before he felt comfortable outlining directly the ways in which the Palo Alto therapists treated symptomatic behavior within the family. By then double bind had made headway "largely because it developed and utilized a new general view of communication." The new field of communication had identified communication and behavior as two sides of one coin. It placed communication within a pattern of events in which there was no single, simple message transacted between two parties, but rather multiple channels of communication modifying one another. Percepts of prior human relationships, defined through ongoing communicative interaction, had formed a vital part of the multivalent pattern, whether these interactions had been congruent or contradictory (Weakland 1976a, 311).

At Palo Alto the therapist's job was always to find a means of intervention that would help them make changes to vicious circles, even including the technique of "prescribing the paradox." A successful outcome of therapy would be instances in when troubling communication patterns dissolved on their own accord as families began to realize why the patterns of communication in which they normally engaged had led them into knots they were unable to untwist. The original model therapists had used to encourage self-correction was of cybernetic circuits, similar to those in a household thermostat, in which the therapist alters reference points in a cybernetic network much like altering different temperature settings in a house. Initially, this model seemed to work despite its simplicity and its evident mechanistic underpinning. As research continued at the MRI after Bateson's departure, therapists found that this model of "first-order cybernetics" was not subtle enough to respond to the variety of issues that they faced. They began to alter their procedures, beginning with the questions that they devised for their clients in their opening interviews with them. Such questions as, "What is the underlying problem?" became changed to, "What is going on in the system of interaction that produces the behavior seen to be the problem?" The therapeutic interview then proceeded to develop a hypothesis, which the family and the therapist created together. In this situation, the therapists at Palo Alto became aware that shared ideas held collectively by family members could not always be verbalized but could be expressed through other communicative means, such as through pantomime.

Increasingly, MRI therapists began to embrace a model of therapy that became known as "a second-order cybernetic approach." Here therapeutic understanding emerges by finding means to describe significant recurring patterns rather than by isolating and analyzing elements within the messaging system. Reframing required collaboration between family members and their therapists (Weakland 1976b, 127). A second-order approach to cybernetics put therapists themselves into the whole circuit of information events in which the therapist was participant rather than observer (Keeney 1982). Even the definition of "solution" to therapeutic problems changed, since part of the new therapy of second-order cybernetics was to ask all parties involved to state what observable behavioral change would signify "success" in therapy. Often this was a difficult question for patients to answer, but from the therapist's standpoint, any small change in awareness could be significant.

The cybernetic events into which therapist and patient now became thrown consisted of the full round of Bateson's cybernetic epistemology, together with self-reflexive perspectives of maps of maps of maps, of punctuations of punctuations, of relations of relations and of differences of differences. "The process [of therapy] is recursive—what one draws, one sees, and what one sees, one draws. The world of cybernetics is [now] the world of mental process in which, as William Blake put it, 'creative imagination proposes outlines of the world'" (Keeney 1982, 157). It introduces a "choreography" into therapy. No longer is there a definite territory, matter, thing or object through which feedback afforded correction. The new therapeutic techniques still gave the therapists a hands-on direction of the whole therapeutic process, but the approach was moving dramatically from its original forms of "pragmatic framing."

Though successive publications from the Palo Alto school found a secure place in the history of psychotherapy, the experimental changes of framing and reframing fostered by the Bateson Group and the MRI began to foster widespread controversy. The experiments with new ideas about "framing" had come along at a time of a rapid growth of professionals in the field of family therapy, a rapid growth that had occurred partly as a result of the successes of the Palo Alto group. At the same time, many of the newcomers had begun their careers in group therapy or social welfare counseling and found the theoretical approach of the Palo Alto school somewhat austere. These professionals remained wary of "theory," while impressed by the fact that the Palo Alto researchers seemed to have discovered a route toward addressing and even managing the most insoluble pathology of mind—that of schizophrenia. They were practitioners whose skills related primarily to their years of undertaking counseling and their experience with a hands-on approach.

The first critical attacks were directed toward the theoretical aspects of logical typing. Family therapy practitioners complained bitterly about the difficulties of understanding Bateson's hierarchy of logical types and were often in despair as to how they would put this idea to practical use. Bateson replied to his complainants that he never meant his discussion of logical types and double binds to be practical tools, but rather a way of thinking, an epistemology that contrasted strongly with the usual techniques of behavior modification. A second set of criticism was even more vigorous. Second-order cybernetics had reframed the relationship between therapist and client by arguing that all worlds of experience are in-formed and self-referential. Family therapists complained that such equality in framing self-reference can verge on runaway when introduced in everyday therapy.

Bateson had supported, even encouraged, the introduction of second-order cybernetics, along with Margaret Mead. His own methodology did consider a second-level pattern, a pattern of generalization about the patterns of therapists engaged in social interactions with their clients, but Bateson's level of logical types did not consider a literal emplacement of a second-order observational level of observers (psychotherapists) undertaking the observation. The approach of second-order cybernetics had come to proceed with one psychotherapist beginning in the usual manner to engage in therapeutic questioning of his or her client, but joined by another psychotherapist as "meta-observer," who observes the therapeutic interview anonymously and out of view, until such point at which the meta-observer decides to enter the therapy directly as commentator on the way the interview was conducted. In some cases, the meta-observer was placed in a room with a one-way mirror placed above the interview room where the therapy was taking place. In other situations, the two levels were enlarged to three levels, as a meta-metasystem. The rationale was to overcome any dualism between the observer and the observed, with each level regarded as complementary to the other levels.

Bateson's self-reflexive levels had merely suggested that therapists have backloops to their inductive style of questioning so that they could more easily deal with the existence of rules propositions and beliefs in the family setting. A backlooping "meta" is a very different sort of lens through which to conduct an enquiry of levels of communicative order than invoking in a physical sense a "higher" and "lower" order of observation of communicative interaction, in which different therapists in different spatial levels address the propositional viewpoints of the observer.

Bateson had also written that the central issues in understanding information were, first, to grasp the contexts in which information exchanges take place, and second, to understand how those contexts relate to each other in an interpretative as well as instructive sense, that is, in relation to the observer's premises about reframing feedback patterns. But the 1970s then ushered in another paradigm of recursive self-reference conjoined to second-order cybernetics, whose origins lay in biology rather than psychology. The paradigm emerged from autopoiesis, or a systemic bootstrapping of recursive paths in living systems through "mutual coordination of mutual coordination." Since autopoiesis emerged from biology, there were challenges as to whether the rationale of its authors was appropriate to human psychotherapy. Of the two authors responsible for creating the concepts of autopoiesis, Humberto Maturana and Francisco Varela, Maturana decided in the affirmative. He believed that his biological notion of autopoiesis was a constructivist model that could be applied to human psychotherapy, while Varela deferred his answer. Meanwhile, the critics suggested that a transposition from the world to biology to human beings and their complex relational networks was a risky venture at the best of times, even more so when these ideas became transposed from theory to everyday practices.

Maturana's ontological cybernetics seemed to change the rules of the therapeutic game. Maturana wrote of the closure of all (biological) cognitive systems to instructive information (Maturana 1984), that is, so-called information transfer from the environment, social or physical. Such information does not and cannot instruct the behavior of the living system, he proposed. Instead, in human circumstances, the therapist must avoid instruction and undertake "languaging" with his or her client instead, for the purposes of consensual coordination of consensual coordinations of behaviors. The theoretical idea is that "languaging" lies less in the domain of instruction than in the domain of systemic operation of the organism as a whole, and suggested that recursive interactions are autopoietically coordinated through the self-corrections of an autopoietic system and not through therapeutic intervention (Maturana and Varela 1992). Thus, the domain of instruction existing between therapist and client is transformed into a domain whose dynamics are those of emotional satisfaction. Change is possible in the interactions between client and therapist only if the client-patient exchanges change the patient's own premises to his/her own emotional satisfaction (Ruiz 1996).

Maturana's notion of "languaging" for autopoietic transform was very different to that of first-order information transfer by the observer-therapist to their clients. Therapists began to feel that Maturana's recursive bootstrapping approach gave too much emphasis to the autonomy of a structure-determined system and too little to the family therapist's own sense of responsibility in initiating any intervention with his patients or clients. It also seemed to permit an "anything goes" type approach to family therapy, a line of argument that made therapists very discomforted. There was still one more looming question. Did Maturana mean that no information in the form of instructive learning was present in the recursion of autopoietic systems? Maturana's virtual erasure of the idea of instructive information seemed to rub out much of Bateson's own approach to learning (Dell 1982, 1985, 6).

A comprehensive discourse began in the journals on the merits of the concept of self-organization in therapeutic practices and "languaging" as therapeutic discourse. The practitioners' response to suggested techniques derived from the autopoietic approach was less kind. They began to ask what happens to the therapist's right to intervene in the therapeutic situation. Second-order cybernetics seemed to put into abeyance the therapist's own skills of knowing and acting, while the client seemed to be put in control. The therapist's ability to undertake family reorganization seemed to have less to do with specific intervention procedures and more to do with evoking aspects of clients' family history, relationships, and relationship rules and myths.

The debates conducted by different authors on Bateson, second-order cybernetic ideas, and autopoiesis should have been enlivening, namely in considering what ways a structure-determined system can be self-interpretative, and to what extent the structure of a self-interpretative system is determined by mutual coordination. Yet debate gave rise to contortions and confusion. Many of the readers of the debates on "epistemology" (Bateson) versus the "ontology" of Maturana published in *Family Process*, a journal that represented the viewpoint of systemic family therapy, became totally intimidated by the complexity of the discussion. Others became disillusioned with alternative constructs (Held and Pols 1985, 516). Many others thought the Bateson group had engaged in overkill. And there was a flurry of scathing dismissals from outraged "pragmatic" therapists. They renamed "epistemology" in ironic fashion as "lenseology" or "epistobabble" and from there on such hostile reaction dampened the free play of ideas in the field (Hoffman 1985).

The confused discourse in systemic family therapy eroded its influence at quite a rapid rate. Within the space of about ten years, systemic family therapy had abandoned Maturana's biological constructivism, as it was known, in favor of its own form of social constructivism (Britt-Krause 2003). The theoretic move drew family therapy into the influence of poststructuralist postmodernism—with its very different set of assumptions about the family as system and a type of constructivism different to that of the Bartlett-Bateson mode (Flaskas and Perlesz 2002, 32).

BATESON AND R. D. LAING

Weakland had summarized Bateson's approach toward schizophrenia as being one of many levels. Nevertheless, an argument for a single-level determinism in schizophrenia, that of genetic determinism, was strongly supported at the time and is still favored by some of the main advocacy groups for patients of schizophrenia. These groups are those who tie the illness of schizophrenia to a medical model and call for appropriate psychiatric care within institutions for mental health. The confrontation between these two alternatives still continues. Those who recognize the validity of a many-level approach point out the fruitfulness of fostering peer group communication about the many facets of their illness. Communication among the afflicted and their families and friends, they believe, is an important aspect in amelioration of pathological effects of schizophrenia and does indeed alter their communicative behavior. Beyond this, devising community or family social group responses to the affliction of schizophrenia is also an important aspect of "cure." A discussion of symptoms has two ameliorative results: It alleviates the pathology, and it becomes a means through which schizophrenics can find a way out of their affliction (MC, 17/07/1967, 1096–27).

The many-level approach is likely to be found today among those favoring alternative medicine. One of the most interesting examples of this is the growth of worldwide movement called Mad Pride. The movement began in Toronto in the early 1990s by a group that launched a Psychiatric Survivor Pride Day in response to local community prejudices toward people with a psychiatric history. The area surrounding the medical health facility was noted at that time for low-income housing, high welfare distribution, and attendant issues of drugs and prostitution. The same group now sponsors a Mad Pride week of events at the same locality, but has become transformed into the bourgeoning entertainment district of Toronto. The aim is for "psychiatric survivors" to reclaim the term "mad" in much the same way as another marginalized group, the homosexual community, has reclaimed the word "gay," a term once used as a slur against them. Mad Pride mixes activism, skill sharing, street theater, and celebration of "mad culture." It is a way of reframing psychiatric illness as normality within a community, rather than letting sufferers' experiences continue to be diagnosed as having an incurable, violent condition with little hope of successful outcome. Its street theater includes pushing a hospital bed along the street outside the once walled and hidden-from-view mental health facility known as the madhouse.

Outside North America, Bateson's approaches achieved greater success, notably in Italy and Great Britain. The Milan School of Psychotherapy disregarded the controversies over first-order and second-order cybernetics and concentrated on Bateson's observations about paradox, and by taking the notion of counter-paradox as its central motif, developed casebook studies of psychotherapy using this methodology (Palazzoli et al. 1978). Not unsurprisingly, the European country most sympathetic to Bateson's view was Great Britain, especially in the writing of the British psychotherapist Ronald Laing. By the end of the 1960s, R. D. Laing was one of the best-known psychiatrists in Great Britain, for even though he himself was of modest origins, he was able to pick up wealthy clients. Laing was not only a practicing psychotherapist but also and above all a writer who excelled in paperback presentation of approaches to madness, illusion, and reality. Early in his writing he had realized how profoundly Bateson's notion of "madness" as an "inner voyage" with its own endogenous dynamics spoke to his (Laing's) own personal experience of mental illness, and should define the therapist's approach to treating pathology (Pickering 2010, 185).

Bateson's approach to psychiatric "treatment" was, in effect, a trial and error experiment with behavior patterns in order to see what works. Even though it was impossible to control those dynamics and guarantee a cure, the psychotherapist or psychiatrist could participate in these dynamics. The task was to assume a relationship with the sufferer, and to latch on to the patient with schizophrenia and with an assumption of the condition as being that of a normal human being undergoing affliction. This was utterly divergent to mainstream practices of the day, which were those of keeping schizophrenic patients isolated in their residences. Isolation of schizophrenics went along with prohibitions against treating any patient in any normal, human, manner. In more than one of his books, Laing cites the horrified response of his colleagues to his acts of talking, sharing with, and befriending his patients, especially schizophrenics. But as the first move toward "cure," Laing immediately proposed a "rumpus room" for patients and staff in order to dispel interactive isolation.

Laing's *Self and Others* (1961) gives Bateson fulsome praise. The work of the Palo Alto research group had tackled the vital issue of how psychological relations between self and other, most particularly those in close relationships to each other, escalate into untenable positions. The Palo Alto group had revolutionized the concept of what is meant by the "environment" of mental illness through taking into account the communicative patterns of patient within the patient's family, and "has already rendered obsolete almost all earlier discussions on the relevance of 'environment' to the origins of schizophrenia" (Laing 1961, 148).

While Bateson had regarded his own development of communicative interaction as being epistemological rather than directly political, Laing's best-known book, *Politics of Experience* (1967), gives double bind a Marxist twist. Any research that challenges basic suppositions about madness must have overtones in the political realm, and Laing thought of double bind as an example of falsified dialectics in Western society. The basis of Western society was competitive exchange, a mode of organization that not only covers values of goods and services but also embodied competitiveness in interpersonal activity within such exchanges. Causal efficacy of interpersonal relations is embodied in, competitive relations of social power, which in turn leads to "falsified dialectics." This situation is akin to double bind in that perceptions of self and other in the interpersonal communications of a society contradict systemic appraisals of social power in society. A victim finds that if the message means A, this invalidates what the victim has learned about the contexts of relationship in which that particular interpersonal message is valid. Or, if the victim chooses B, then the victim must deny his or her perceptions about the systemic significance of social power, the "truth" value of which the victim has learned at some prior stage. The metaconditions that satisfy both A and B are absent (and cannot be discussed), leaving the victim to oscillate between either choice A or choice B. Since each choice facing the victim proposes, but does not satisfy the other, any choice results in invalidating the process of choosing alternatives—as the victim oscillates between the alternatives levels of personal and systemic significance.

Bateson thought of double bind in terms of rebutting the false dialectics of game theory wherever game theory drew upon Arrow's utility theories about competitive exchange. Yet both game theory and utility theories are still prized by conservative politicians in the Western world, so that double bind can indeed be visualized as a counternarrative that extends far beyond the original case study into the political realm. In a technical sense, Bateson, unlike Foucault, avoided being directly political. Bateson's attack was on false value systems compared with Foucault's attack in his history of madness in Western Europe, which was a direct attack on the activity of the state (Foucault 1988).

Bateson argued that his epistemological approach could also be valid in studies of the binding forms of addiction. His "Cybernetics of Self" (2000, 309–337) was a study of addiction to alcohol in which the false value system examined occurred among individuals who believed that they were in control of their addictive habit. He noted that the alcoholic is in constant warfare with the bottle through his or her drinking habit. It is a warfare that evokes repetitive pain and despair, pride and good intentions, and the addict consistently fails. The first step in recovery is to acknowledge that, as addict, he or she is in a situation out of control, but as Alcoholics Anonymous stresses, there is another expression of control, a meta-expression derived from being among a group of people with similar conditions. This can bring hope and change the alcoholic's view about a dawning of new possibilities. That hope emerges from understanding how to let the condition be (Harries-Jones 1995, 38–41). Chapter 9 will discuss when humans take on nature and when they treat global climate change in terms of utility theories, how their false dialectics of control create conditions that resemble double bind.

INTERLUDE

From Cultural Structures to Structure in Ecology

As a postindustrial thinker, Bateson saw communication as the primary catalyst for human organization and in a wider dimension, a catalyst of order in living systems of which humanity is a part. His epistemology of communication lay in between the fallacies of upward reductionism of the notion of structure first expressed by Whitehead (and discussed in *Naven*) and the fallacies of downward reductionism of structural complexity evident in the protagonists of behaviorism, the game-playing RANDites and military establishment, and in Artificial Intelligence.

His perspective would probably have become much better known had he cast it in the political terms employed by Laing, Deleuze, Guattari, and Foucault, but he had an aversion to that form of format. Félix Guattari's *Three Ecologies* (2000) mixed Bateson's ideas together with Guattari's own objections to capitalism and postcapitalism, but it fails to realize that Bateson's counternarrative is at a different level of inquiry from the primal screams of political and economic objection. Bateson pointed his readers away from epistemologies derived from technical order, from the material structure of politics and economics, to the more abstract "structures" of pattern formation to be found in ecology and evolution.

In this transcontextual shift, Bateson redefined the notion of "structure" in keeping with notion of feedback patterns playing upon transform in coded rules of ecological interaction. Depiction of ecological structure was more like a "context of observation" at selected interfaces than an underlying determinate form. The processes of ecological feedback could not be described in real time, yet an observer could pick up variety in pattern transformation and indicate aspects of pattern survival. By means of a hub-and-spoke approach, a glimpse of structure could indicate how technological development might impinge on the stability of evolutionary and ecological order and lead on to a discussion of human survival. Overall, his concerns about structure were linked to human survival in rapidly changing ecological circumstances.

Nevertheless, there were contours of Bateson's thought that overlapped with that of twentieth-century poststructural critiques. Like Deleuze and other poststructuralists, Bateson rejects dualism and seeks to liberate thinking from binary opposition and open it up to multiplicity. Bateson's treatment of "bind" and "double bind" replaces the tight constraints of contradiction encased in Hegelian and Marxist dialectical oppositions (i.e., terms of negation and negation of negation) by the communicative concepts of feedback and feedforward. Contrasts such as inversion, rotation, and the like were thereby abstracted from the confines of Hegel's narrow logic of opposition and considered instead as contrasts of perceptual gestalt. Deleuze opposition to Hegel had stemmed from similar objections to the absolutism that he found in Hegel's notion of contradiction. Hegel had considered "difference" only in terms of "absolute maximum of difference" (Currie 2004, 64). Bateson evidently agrees with Derrida that it is appropriate to abandon the dominance of classical logic when it comes to understanding communication.

As we have already seen, Bateson's perceptual processes of difference owe little to the logical truths of induction and deduction but are formalized around a propositional ordering of differences among and between relations in a system. This leads him to favor of the use of metaphor in his propositional logic. As with other poststructuralists, Bateson believed that metaphors are the means through which all qualitative thinking begins and through juxtaposition and analogy enables "difference" to be married to equivalence and balance. His notion of logical types, a temporal logic of "in-between," enables him to move from forms of distinction framed as categorical oppositions of truth and fact, and which are always difficult to bridge, to difference in ecological order, where no categorical dialectics exists and distinctions of identity-alterity, that is, relational distinctions, are perceptual and analogical.[1]

For Bateson, the ordering in communication requires not only substitution of classical logic but also that "difference" within interactive communication of living systems cannot be explained through categories of language. At the first-order level response, meaningful content may arise from sound patterns or distinguishable rhythms, or indeed the ordering of words, but meaning also arises from second-order response, which is meta to sounds, rhythms, or other media. This second-order response creates contextual meaning. Context in this sense is metaphysical, residing in "the form of the form," which is not to be confused with meanings derived from linguistic categories. As he put it: "the name of the name" for the concept of difference is not in itself a name.

Difference patterns enable correlation with any event that is "news" or "variety." And "difference" is a primary factor in perception. "Difference" is not to be taken as "the other side of the coin" to "sameness," since difference is always the means for generating perception of pattern. Note that Derrida takes "sameness" and not "difference" to be primary; but Bateson argues that "sameness" always yields the question, "How much sameness?" a question that is always answered in terms of quantification, and not in terms of perceptive contrast (MC, 01/06/1967, 1519–1b).

Bateson's counternarrative stressed qualitative distinctions over quantitative distinction, recursive ordering over linear ordering, multilevel patterns over single level patterns, a logic that is temporal rather than static logic, and communicative forms that are primarily analog rather than digital in their coding. Given these requirements, it was always difficult for him to choose clear examples of metarelation connected with his notion of "logical types." In his final years, he tried abandoning Russell for the mathematics of group theory, but never found a topology that satisfied him. As we shall see in Chapter 7, the topology that came closest to his requirements for "heterarchy" was developed by his friend Heinz von Foerster. Nevertheless, he succeeded in using Russell to show that fixity of rules in communicative interaction often led to paradoxical interactions, a characteristic in pathological families, such as in family communication of schizophrenics. In "normal" families, formal patterns of communicative interactions were *heterarchical*, that is to say a multilevel ordering in communication of recursive events. The task of psychotherapeutic invention in pathological situations was to move from fixity to flexibility in family interaction.

Bateson viewed with concern the rapid transition occurring in the technical order as it switched from analog code to the digital code of numerically controlled machines. This switch, from the 1960s onward, affected cybernetics itself. He feared the growing importance of "the mechanists" and the cybernetic engineers of Artificial Intelligence (the "mechy-machs" as he once called them). As another critic expressed the situation, Artificial Intelligence had taken an inanimate object, the computer, as the prime metaphor, and it seemed to propose that humanity and nature are a unity as a result of both being ontological artifacts complementary to each other (Bolter 1984, 219–220). Not only did this violate the fundaments of Bateson's position that humanity is part of nature, but Artificial Intelligence also seemed bent on the task of imitating nature in order to improve upon it, a prospect that Bateson regarded as a total disaster: "Recently with the terribly rapid development of electronic devices and especially digital computers, the trend has been to a greater hubris, a tendency to see man as potentially 'captain' not only of his soul, but of all that surrounds him (including the souls of other men. It becomes important therefore to re-examine the unit of mental process, purpose, ecological change and biological evolution. What is part of what?" (MC, 15/03/1973, 824–2).

Other poststructuralists who shared Bateson's ideas, such as Paul Virilio, declared that the development of a purely statistical notion of information developed, as we have seen, through the strategic necessities of intelligence during World War II, had created situations that were pushing cybernetics away from democratic institutions. Virilio called this trend the creation of "information bombs" (Virilio 1995, 133): "The first deterrence, nuclear deterrence, is presently being superseded by the second deterrence: a type of deterrence based on what I call 'the information bomb' associated with the new weaponry of information and communications technologies, and along with the 'information bomb,' the embodiment of virtual reality." Virilio and Bateson highlighted game playing as the central metaphor in this situation. The cascade of information bombs derives from gamblers who cannot live without chance and who rapidly become addicted to game playing, Virilio states. To play means to choose between two realities—the virtual and the real. As these gamblers move in and out of the real world and into the virtual world on which they are hooked, they undergo a constant switch in framing of their perceptions. At the same time, their virtual world develops into an expansive space far beyond that to which Alan Turing referred in his celebrated paper "Computing Machinery and Intelligence" (Turing 1950).

Turing's prophetic account of Artificial Intelligence had set a huge trap for the unwary. Turing's "imitation game" assumed a complementarity between the human mind and the electronic activity of the machine that could never exist. As Joseph Weizenbaum, along with Virilio, states, there are a number of correspondences to be made between the iterations of a computer loops and that of recursiveness in human thinking, but there is no complementarity between the symbolic manipulation of electronic space, and the sociopolitical-cultural space in which human discourse takes place. Computer "thought" is a sequence of operations of fetch-and-execute cycles of events. The symbols that a computer manipulates have no reality behind them. Given a certain pattern of bits in input, the machine will produce another pattern of bits as output. Since electronic space is itself an artifact, a constructed space, the path from input to output has to function in a thoroughly predictable way, a path which is totally at variance with human decision making outside electronic space (Weizenbaum 1976).

The detachment of Artificial intelligence from environment, social or natural, alarmed Bateson, for by creating a realm where the manipulation of formal rules upon contentless symbols results in a product for machine use, A.I. ends up with an embodiment of the world that the logician would like it to be. It becomes a highly digitized world, rather than what it is, a mixed analog/digital world in which analog is dominant in nature. He begins an unpublished article about this prospect in one of his Notebooks (*Notebook* 45/1971).

The critic J. D. Bolter argues the same case. The computer has conquered the disorder of the natural world by harnessing the hierarchical principles of symbolic logic to computer circuits in such a way that the circuits designed the principles of hierarchical order themselves (Bolter 1984, 73, 187). As human experience comes under the command and control of AI, the tasks of imitation proceed apace, and the computer programmer begins more and more to rearrange human experience to the meet the logic and symbols of electronic space. By overlaying sociopolitical "real" space with electronic space, grand programmers emerge as "gods" who assume the position of creator-deities making the world many times over again, rearranging its elements to suit each new digital program of their creation. Yet, Bateson believes that all relationships and other interactions of organism and environment were forged through coding duality, both analog and digital (see Chapter 8).

In Bateson's mind there is a recognizable distinction between the earlier days of cybernetics, the days of Norbert Wiener when cybernetic ideas had reference to "real" events in human or mechanical existence, and a later stage when these same ideas became embodied in the virtual realities of Artificial Intelligence. What is most important, the AI path could never deal with any whole, nor did that path seek to deal with wholes. The rules of hierarchical ordering which underlay all AI were inappropriate for this task.

As we shall see in Part II, he moves on to biology and ecology where despite the discovery of the pattern of the genetic code, the "central dogma" of biological organization proposed that that the gene within the genome is a physical entity; it also proposed that the genome is the template for all biological organization. Bateson's view was that the materialist template that molecular biology had encased in its central dogma supported dualism in biological organization. Moreover, biologists continued to insist on rigid isolation of interior genetic processes from exterior adaptation. This meant that the rules of change in the two instances were very separate. One explanation of change stemmed from the pressure of *exterior effects* on an organism (Darwin's natural selection); another explanation of change was that of mutation in *interior effects* (genetics). Bateson believed that the rules of biological organization were never additive (exterior rules plus interior rules) in this manner. His view was that all evolutionary changes were associative, interactive, mutually supportive, network-like and fractionating, thus creating multiple levels of response. Further, evolutionary change was always an event involving coevolution (see Chapter 7), not simply "force" or pressure exerted by environment on individuals: "I suspect that the word 'pressure' is used by Darwinians and such to avoid another word which they think would be 'unscientific.' Do you know what that word is? The word is 'preference.' But then if you let that word in, you have to let in the notion that organism-plus-environment is a system which the vulgar might call a mind" (MC, 30/05/74, 1400–9b).

Peculiarly, Bateson found congruence with Alan Turing, for in a paper written *before* the Watson and Crick uncovering of the helical pattern of DNA, Turing posited that the process by which identical cells in a developing organism differentiate into the various cells that make up the organism's adult form (morphogenesis) was the most appropriate approach to fundamental questions about life (Turing 1952). Turing stated that the puzzling question of how the undifferentiated cells in an early embryo begin to specialize, with one becoming a bone cell and a second a blood cell, occurs through self-organization. The same is true of all the shapes and surface patterns we see in the huge diversity of plants, insects, and animals. He theorized that cells change shape because chemicals in an embryo react with each other and diffuse across space, and he provided a mathematical model of the chemical processes underlying cell differentiation. To develop variety of patterns and different shapes, he argued, these chemical reactions need an inhibitory agent, to suppress the reaction, and an excitatory agent, to activate the reaction. Then chemical reaction, diffused across an embryo, will create patterns of chemically different cells. Developmental biologists adopted aspects of Turing's thesis of morphogenesis, as we shall see in Chapter 7.

Turing's paper lists six different patterns that could arise from this model (Evans-Pughe 2014). His paper took sixty years to be confirmed. In March 2014, scientists at Brandeis University and the University of Pittsburgh published a study offering experimental evidence confirming Turing's theory. In August 2014, a team of scientists led by James Sharpe from the Centre for Genomic Regulation in Barcelona showed that the way fingers and toes form is orchestrated by three molecules in the same manner as Turing's theory had suggested. It also helps explain all sorts of biological phenomena, from the pigmentation of seashells to the shapes of flowers and leaves. Bateson also believed that the study of morphogenesis offers a better chance of finding answers for many fundamental questions, rather than attributing all biological change to coding in genetic templates. The major difference in Bateson's discussion of morphogenesis from that of Turing's biochemical field is to ensure that transformation of the form of the form in morphogenesis is shown to play out through differences occurring from *outside* in the immediate environment conjoining together with pattern of redundancies occurring *inside* the system: "Wherever difference may be 'it' can trigger the degradation of energy without itself self-supplying energy. In other words difference is a truly psychological concept . . . [and] so long as we talk about difference do not pretend that difference is somehow physical" (CAF 308/1971).[2]

A major issue in the forthcoming chapters is whether Turing's purely physical rendition of biochemical interactions in morphogenesis should be replaced by Bateson's realization of difference as creating an interpretative order so that order is recognized as being a metalevel to physical biochemical interaction. This is the premise of biosemiotics discussed in Chapter 8, and it is made even more enticing by the fact that Turing recognized the role of rhythm in creating pattern oscillations, making the drops communicate so strongly that they suppress the oscillations and adopt a periodic spatial pattern in which some drops are chemically off and others on. Are these evident rhythms to be explained simply as the base of biochemical phenomena, or are they the result of communication activity in which live agents are responding to difference? That "which the vulgar might call a mind"?

PART II

NATURE'S BALANCE

6

PATTERN AND PROCESS

Bateson decided to leave his Palo Alto colleagues in 1965 in order to take up research on dolphins. He was fed up with criticism around the double bind research and the debates he continued to have with colleagues over the direction of their active work in psychotherapy. For most people, a switch from psychotherapy to study the "mind" of cetaceans would have meant moving into an entirely different field of scientific enquiry. Yet, just as his move into psychotherapy was initially prompted by his professional work as an anthropologist and led to his starting to undertake an ethnography of the Veterans Administration hospital, so too did his move to study dolphins arise out of existing professional interests, in this case both communication studies and biology. As our earlier chapters have indicated, Bateson had always been a biologist, since his undergraduate days at Cambridge University, and much of his work both in anthropology and on communication had been prompted by biological ideas. As well, his social-psychological papers on learning that he had written before he joined the military had prompted ideas about a wider notion of learning, with an emphasis on habits, apperception and recognition of context, all of which could be its own framework of enquiry, not only in the human world, but in the wider ecological world as well.

In 1962, a year before Bateson left Palo Alto, Rachel Carson published *Silent Spring*. The central themes of this most influential scientific study concerned widespread use of pesticides, especially of DDT and the way in which pesticides and herbicides worked their way through plant and insects in the form of bioaccumulation. Bioaccumulation is a process whereby a toxic substance is absorbed by the body at a rate faster than it is lost, and where biological magnification occurs as the substance increases concentration along a food chain. Bioaccumulation can kill birds, but it can also cause genetic mutations and cancer in humans. The backdrop to Carson's book was the widespread public concern about radioactive fallout as a result of the aboveground testing of nuclear weapons distributing radioactivity by wind and water upon and into living creatures. Secrecy surrounding the fallout problem had begun to lift in 1954, enabling the scientific community to study the extent of environmental degradation and contamination caused by nuclear weapons tests. In 1957, the Pugwash Conference addressed the control of nuclear weapons, and a 1958 petition to the United Nations, initiated by the Nobel Prize–winning chemist Linus Pauling, called for an end to nuclear weapons testing. Pauling pointed to the myriad biological threats associated with carbon-14 and other human-made radioactive isotopes, such as strontium-90, iodine-31, cesium-137, and carbon-14. Once lodged in all the tissues in the human body, these isotopes become part of its bodily composition. It was at this point that, perhaps for the first time, the public became aware that the whole world's population of human and nonhuman animals shared a common environmental fate.

There is little doubt that the close relationship between nuclear testing, nuclear fallout, and the arms race was a prime factor drawing Bateson toward writing about ecology. As prior chapters have shown, he had studied the pattern of arms races and after the war had joined a committee of scientists that sought to understand the effects of nuclear weaponry on military defense strategies and international politics (Bateson 1946a, 1946b), a group that included Robert Oppenheimer, administrative head of the Manhattan Project.

In 1958, a segment of American scientists began organizing a Committee for Nuclear Information with an aim to strip the secrecy away from the weapons program and warn their fellow citizens of the dangers—at the very least to stop nuclear testing above ground. In 1963, it renamed itself the Committee for Environmental Information and launched a new magazine, *Environment*. A prominent leader was Barry Commoner, who later was among the group of scientists whom Gregory Bateson called together at his 1968 conference in Burg Wartenstein, Austria, on the "Effects of Conscious Purpose on Human Adaptation." Edited by his daughter, Mary Catherine Bateson, the book of the conference appeared as *Our Own Metaphor* four years later (M. C. Bateson 1972).

Commoner and other participants of the Burg Wartenstein conference favored a call for political action on behalf of the new environmentalism, but Bateson rejected this, considering such a move premature. He was more resolute in 1979, when he resigned from the board of regents of the University of California. The board had voted to support increasing research on nuclear weaponry in the university. He published his objections under the title "Nuclear Armament as Epistemological Error" (Bateson 1979b). His resignation letter repeats some of the themes of earlier papers on armaments races and zero-sum games, notably that any attempt to initiate "trust" between the United States and the Soviet Union would be much more likely to bring the nuclear arms race to an end than the constant tit-for-tat approach to military capacity of nuclear weapons. Bateson proved to be correct. The talks in Iceland between Gorbachev and President Ronald Reagan did indeed initiate trust between the two superpowers and did prove to be the beginning of the end of the Cold War. All this was to happen after Bateson's death.

The real importance of *Silent Spring*, as Bateson realized, lay in the way in which it revealed profound interconnections both within nature and between nature and society—in effect, the interrelatedness of the web of life, or as he expressed it, that the unit of survival was organism plus environment. As usual, Bateson's position reached into the realm of epistemology. The application of modern physics through nuclear research now threatened widespread death and the destruction of the human species. Could the new science of ecology, in which so many had placed their hopes, prove any better if that science embraced the same framework of thinking about the preeminence of the material forces and applied these to the study of nature? Many of the new books and papers that called themselves "environmental" concerned themselves almost entirely with issues of biomass and bioenergy, exhibiting little change in conception about the nature of living systems from the mechanistic determinism that dominated much of physics and chemistry.

This was one epistemological issue. Another was the rhetoric of some ecologists. In the 1960s, it had become the fashionable for some ecologists to drop out. Such rhetoric would achieve little if their research curled itself up in a world apart from natural science. It seemed to Bateson that if many who now labeled themselves "ecologists" were to follow this path, it would take them into a physical and intellectual wilderness. By Earth Day 1970, a generally acknowledged marker for modern environmentalism in the United States, there were indeed sufficient new books and papers calling themselves "environmental" or "ecological" to herald a whole new subject of enquiry. Ecologists had emerged in this new age of atomic anxiety, as the "guardians of fragile life" (Worster 1994, 340, 343). Yet the methodology of ecological studies, from Bateson's point of view, still paid little attention to animal learning, and did not incorporate such study into an enquiry on evolution and ecosystem adaptation. Ecology as a field of enquiry remained within the same framework of thinking as that which he had examined twenty-five years before when he had criticized Evelyn Hutchinson's precybernetic approach. If the so-called mystery of ecological unity was to be investigated, the major focus should be one of how parts fit into a holistic order, and vice versa, how holistic order is contained in the development of parts. Such an investigation would require an approach very different from the reduction of ecological order to physical events and then experimenting with their productive economic utility.

The literature of cybernetics clearly was a starting point for this new venture. Yet it was not sufficient to deal only with cybernetic processes in ecological order. An additional investigation was required, a second-order investigation of the human perception of, and response to, the awareness of cybernetic processes in ecological order. Another path was to insist that any observer of biological systems is always engaged in *participant investigation*. Living systems are recursive systems in a way in which physical systems are not, which means that in the case of errors in perception on the part of observers, these errors will always come around to stab the observer in the back. A broader task was to ensure that the object/subject relation in ecology became revised (see Chapter 8).

THE MUDDLE OVER BIOINFORMATION

A physical definition of information continued to be dominant in ecology. The conditions prevailing in Bateson's time were such that hardline scientists would not consider biology as a science because its principles of order did not seem to correspond with the laws of thermodynamics. Many scientists in physics and chemistry classified themselves as "hard scientists" by virtue of the fact that their work always followed the constraints of known natural laws. Among the most important of these were the two laws of thermodynamics, the second law of which concerned the dynamics of a heat source and its movement toward entropy. In its broadest definition, thermodynamic entropy is the amount of energy coming from the sun to the earth balanced by the energy reradiated from Earth to space. Incoming energy creates disequilibria in the atmosphere, lithosphere and hydrosphere, together with reservoirs of latent heat and mechanical energy yielding a statistical distribution of short-term randomness and long-term coherence. In a narrower definition, the law refers to a rate of flux from a heat source moving from heterogeneous to homogeneous order.

Wiener's original definition of biological negentropy, or feedback, was devised in relation to thermodynamic entropy as "islands of order" in a sea of flux that overall was moving toward homogeneity, just as the second law of thermodynamics requires. Yet the link to the second law was tenuous, since information was not a gas, nor was it operative within a closed system, so Weiner's "negentropy" might have been useful in establishing cybernetic feedback within the tradition of scientific thought, but otherwise created considerable confusion. Nevertheless, information theory was used to support a hypothesis that "ecological diversity begets stability." Its first uses in ecology matched diversity together with negentropy—the idea that through variety and redundancy, living systems can build up organization for periods of time, which moves living systems against the general flux of energetic flows, from order to homogeneity.

In 1973, an ecological mathematician, Robert May, advanced a thesis that the existence of increased connections among the members of a biological system (that is, greater redundancy, greater biodiversity) is *more likely* to degrade the energy flows of an ecosystem than to restore its stability. It was a thesis that, if true, would have brought standard interpretations of information theory in ecology to their knees. May's supposed mathematical proof of the contrary condition resulted in an even greater quandary about information theory and cybernetics among ecologists.

The notion that diversity begets stability was a proposition about how redundancy of pathways among the compartments of an ecosystem affords possibilities for rerouting energy transfers so that variety retains ecosystem equilibrium through diversity of channels—otherwise an ecosystem would become quickly degraded (May 1973). This proposition treated "redundancies" in ecological systems in the same sense that a communications engineer would treat material or energetic redundancies in a telecommunications system, and not as a web of interlaced multiplex patterns of variety in *living* forms. Years later, Robert May's mathematical proof was shown to be very limited in its range of application and his account became disregarded.[1]

Ilya Prigogine's elaboration of "dissipative" energy systems, that is systems which exist far from energy equilibrium, also helped to rescue the notion of ecosystems from May's mathematically derived quandary, and restore the validity of the propositions of diversity begetting stability. Prigogine showed that dissipative systems—of which ecological systems were a case—do not behave in their dynamics like mechanical systems (Capra 1996). In recent years, there has been another considerable shift in attitude regarding exactly what probabilities information measures. Summarizing this shift, Robert Ulanowicz states: "This change has been away from the idea that probabilities measure our [the observer's] ignorance about a deterministic situation, toward the notion that they reflect *indeterminacy inherent in the process itself*" (Ulanowicz 1997, 65).[2] The general idea has gained ground that if information has a relation with entropy, it is because information refers to *system constraints*, and those constraints, in turn, impart order and pattern to a system as a whole, instead of being particular states of observer's knowledge under uncertainty. This is more in line with Bateson's thinking, except that one would have to add that systems constraints are ordered by the activities of living systems.

BIOENTROPY

The two-decade confusion had resulted in a chilling effect and many ecologists simply avoided the whole question of information in their appraisal of ecosystems. When Norbert Weiner defined information feedback as "negentropy," he was careful to indicate that "negentropy" was a process that only locally reversed the ultimate drift toward entropy in all energetic systems, and that this "local" reversal of energetic drift sustained biological order in the short term. Since the disorder of a system is measured by its entropy (represented as a dimensionless quantity), the information of a message was defined by the dimensionless negative entropy of the system, that is, by the negative of its probability distribution. Bateson agreed, of course, that organic systems require a positive energy budget in order to survive, but he believed that the subsequent expression of positive energy budgets in relation to the capacity for work did not sufficiently acknowledge how the ordering components of information were inherent in ecosystems. In other words, both Weiner and Shannon, in defining information, had met the requirements of a physical research program where a message was treated like any other element of a physical system, which, of course, Bateson believed was a fallacy.

As Bateson realized, Shannon's formula does not envisage entropy as being in any way related to information as a nonphysical or metaphysical aspect of matter. Physicists might simply disregard that issue. In biology, this disregard is somewhat more serious. Cybernetics had proposed that information is an inverse function of *disorder* over a limited period, which affects part-whole relationships of living systems. This would have had little or no consequence if there had been two registers of information kept separate and used differently by biology and ecology on one hand, and physics on the other. Then the fact that they both labeled these registers as *"information"* would have been an unfortunate historical accident. But the contrary occurred. Leading ecologists, such as Howard T. Odum, stuck with their appraisal of living organization *solely* in terms of its physical properties of biomass and energy and used conventional definitions tying ecosystem order to thermodynamic order. Thus ecosystem = energy = capacity for work. Both Howard and his brother, Eugene Odum, held that "ecology must develop a unified theory of the ecosystem, described in precise mathematical and statistical terms, if the field is to be of any practical value" (Worster 1994, 362). Eugene concerned himself almost entirely with the analysis of biomass and bioenergy, while Howard examined the unity of nature through the lens of thermodynamics and the maximum power principle (E. Odum 1970; H. Odum 1971).

The academic world seemed to accept both of the brothers' models as valid preliminaries for the ecomanagement of nature. The influence of the two brothers stood out in this new field of ecology and environmentalism, but they troubled Bateson. On one matter alone there was a large difference of opinion: Measurements of bioenergetics were always calculated through rules of addition and subtraction. Yet an information-based ecology deals with the *budgeting* of pathways in networks of information through the ecosystem; thus, change could only be reckoned through a ramifying *fractionation of pathways*, rather than addition or subtraction. Thus ecological depletion of information creates loss of overall shape and form, as when hunger through loss of food depletes the whole body; but one does not lose an arm or a leg. Shannon's physical conception of information neglected specificity. That is to say Shannon's concepts of information had nothing to say about patterns of temporal concordance other than those arising from stochastic probability (Bateson 2000, 466–467). Yet, sequences of redundancy define the concordance of biological and ecological information through phases of specific timing. Timing is an all-important point that will be considered further in Chapter 7.

Both Odum brothers spoke of the unity of nature, and the necessity for a holistic approach, but both kept within strict premises of scientific materialism. Their characterization of holism had to be challenged. To Bateson's way of thinking, it had become too easy for modern science to continue to treat the biosphere as it had treated any other "mechanism." The "soft science" of biology had it right when it noted that in biological flux, the rhythm and form of metabolism rates are different in kind from physical flux. Living molecules do not necessarily display a direct response to changed conditions of physical flux. Within their thresholds of tolerance of heat or cold, organisms have the ability to regulate their thermal, biochemical, or other metabolic activities and compensate for changes that result in any imbalance to their activities of reproduction, growth, and mobility. Appropriate information derives from both sensation and experience in all of these activities, and is vital within the physical thresholds of heat/cold tolerance.

Epistemological issues had to be approached decisively if the new science in which so many had begun to place hope was to prove better than its forebears. The first issue is to show how biological flux is different in kind from physical flux, and given their thresholds of tolerance, how living molecules do not necessarily display a direct response to changed conditions of physical flux, as the older discussion of the relation of information to physical entropy assumed. Second, a new epistemology requires a major investigation of the "mystery" of ecological unity that should ascertain the processes through which information enables parts fitting into a holistic order, and vice versa, how information of holistic order is contained in the development of parts.

Bateson took a radical position. He proposed a new term, "bioentropy," that focuses on the ability of organisms to create pattern from noise in any informational context. As a reconfigured notion of entropy, bioentropy could be examined as an analogue of flexibility in a field of multiplicative information patterns. He referred to bioentropy as part of an "economics of flexibility" necessary for any organism subject to evolutionary change. From this, he began to develop the notion of ecosystems in ecology as a contextual ordering of flexibility. The following passage suggests one of Bateson's formulations:

> It is worth noting here that (biological) flexibility is to specialization as entropy is to negentropy. Flexibility may be defined as *uncommitted potentiality for change*. A telephone exchange exhibits maximum negentropy, maximum specialization, maximum information load, and maximum rigidity when so many of its circuits are in use that one more call would probably jam the system. It exhibits maximum entropy and maximum flexibility when none of its pathways are committed. (In this particular example, the state of non-use is not a committed state.) It will be noted that the budget of flexibility is fractionating (not subtractive, as is a budget of money or energy). (Bateson 2000, 505)

Thus "flexibility" of biological order rests within its communicative structure. Success in surviving lies in avoiding maximal rigidity by creating new patterns from noise or ensuring that existing patterns do not become locked into situations not easily changed.

His point is more readily grasped if we take information in an ecological setting to be patterned quite differently from classic Shannon formulations, where messages are sent and received along single channels. As mentioned earlier, Bateson portrayed ecosystems as a field of information where messages correlate and trigger responses through entire feedback circuits of ecosystems, each with varying levels of feedback. Over time, feedback among a whole array of levels enables ecosystems to maintain their organization. When information triggers allocation of energy flux, then this not only touches upon particular organisms and particular species, but it also affects the composition and persistence of populations of plant and animal species in relation to each other throughout the whole ecosystem. Any ecosystem has a budget that is related to its flexibility, or ability to undergo change.

There always has to be a medium for flux that is material, but at another level, the ordered use of flux relates to the way in which pattern is recognized in relation to noise. If we take information in an ecological setting to be "variety," as the cyberneticist Ross Ashby originally used the term, then a continually changing natural environment yields "variety." That is to say, information is not simply sent and received, as in the classic Shannon formulations which more or less mimicked the way information and response used to travel through a central switchboard in an urban telephone center, but correlates responses through multiple feedback controls at several different levels of the "circuit" of an ecosystem. This triggering of response to conditions in an environmental "field" is a result of organisms deriving patterns from information messages and trigger responses to those patterns. The process enables individual organisms to maintain their organization in the "variety" of an environment.

Bateson also uses the term "bioentropy" to distinguish how a process of degradation, through the dynamics of information, is also distinctive from degraded flux of physical entropy. Ecosystems are able to sustain themselves in the presence of conditions of the ongoing degradation of their thermodynamic environment in a way that physical systems cannot. The way they manage to escape the constraints of the physical world of thermodynamics—at least for interim periods of time—is by recycling pollution or excretion so that excretion from one sort of living system is food for another. In the short term, all living systems are antientropic in terms of thermodynamics, but raises the question of how short is "short," that to say is whether ecological degradation in times of turbulent change reveals itself through informational disorganization *before* thermodynamic or energy breakdown occurs. Bateson thought so.

Bateson's answer is that while all organic systems require a positive energy budget in order to survive, yet when it comes to ecological degradation, the primary stresses will be on the formal properties of the organization of living systems, patterns of information. Degradation of information in biotic systems, for example through systemic loss of biota, leads to maximal use of all available circuits of change in a highly patterned feedback system. When stresses are "about to be jammed or seized up" in a network of information circuits, adaptive variety through reorganization of feedback in network connectivity can no longer take place (Bateson 2000, 505). The loss of differences makes a difference to the entire field.

Bioentropy, then, radically changes the scaling of degradation in relation to biotic organization, and shows that—relative to the timing of degradation of physical systems—the conditions for bioentropy in living systems are different from physical flux. Bioentropy begins with noise and little pattern, so this initial condition of "uncommitted potentiality for change" defines maximum flexibility for any system. Normally, as biotic systems develop pattern from noise and the continued input of pattern leads to variety and highly ordered and specific responses, they are constrained by stochastic probability, but more important by the pattern of their interconnectivity, which includes coevolution. We will consider coevolution in other chapters. In the final chapter, we will take a look at Bateson's proposition about bioentropy in times of current depletion of the population of butterflies and the example of colony collapse of honeybees.

In cases where there is a wide variation in response to conditions of physical flux, symptoms of this are visible to humans. For example, animal dormancy is widespread and so is dormancy in plants that grow in harsh environmental conditions, the desert landscape revealing on occasion plants that can lay dormant underground for many years, then with a touch of rain suddenly emerge, flower, reproduce, and lie dormant again, all within a remarkably short period of time.

The way that information triggers allocation of energy flux is, however, a systemic condition which touches not only upon particular species, but also affects the composition and persistence of populations of plant and animal species in relation to each other. This pattern is less visible, difficult to visualize and even more difficult to ascertain. All living forms in their mutual interactions exhibit expectancy—the most evident examples occurring in mating response as behavior evoked in sexual reproduction—but all expectancies are subject to surprise, namely conditions of change.

The notion of bioentropy refers to the fact that conditions of life require semantic or semiotic capacity, an ability to perceive and to translate their perception into meanings, even in quite minimal ways. Change in interactions occur both in anticipation of, and in response to, increased ecosystem stress. Stressed ecosystems provide a context for change, as all biological information, like human messaging and communication, requires contextual interpretation. In this continually changing natural environment, living organisms develop different patterns of response to information. Bateson gave examples of how response rates of selection change with time, and thus how an adaptation may be lost because the response rate may be divergent from the rate at the time when the initial pattern of response rate was acquired. No wonder Bateson argued that the most appropriate definition of survival is survival of the most flexible. The strategy for survival of a species may be immediately governed by contingency but is continually being tested in terms of longer time spans, larger gestalten, and unpredictable change that cannot be foreseen, so that the fight is not only to the strong and well adapted but is also to the most flexible, since the ability of that organism which is at one moment prospering in a given environment may find great difficulties to survive under change and adverse conditions. This brings up the question of natural selection, which since the mid–nineteenth century has usually be interpreted as a force for change. But this interpretation is one the nineteenth century's greatest mistakes, says Bateson:

> Natural selection is a force for staying put, for going on with the same dance [of relationships] that you were in before, not for inventing new dances. . . . What you have got to do is to change in such a way that the system of changing has a certain steadiness, a certain balance, equilibrium . . . [and it] may be a very complicated one. Evolution is essentially a vast operation of interlocking changes, every particular change being an effort to make change unnecessary, to keep something constant. (Bateson 1991, 276)

nATURAL SELECTION

Bioentropy introduces concepts of mutuality and interdependency as core processes of coevolution. By implication the notion of bioentropy challenges the Darwinian notion of adaptation through natural selection. Darwinian selection privileges competitive exclusion through competition from other creatures, or the external stress of environment, or both combined. Bateson argued that over the years that Darwinists reduced the complex variety of activities involved in survival, reproduction, and species interactions, to a uniform interpretative rule, that of "natural selection." They then turn this supposedly universal characteristic of response to stress into the status of a natural law.

Nevertheless, later generations of Darwinists have accepted that evolutionary change is not, as Darwin originally imagined it, one lengthy time span of continuous fine-grained adaptations, but that it is one in which evolution undergoes "punctuated equilibrium." The phrase identifies continuity in time spans of fine-grained adaptations which are suddenly punctuated by fairly rapid change—such as that which occurred with the disappearance of dinosaurs and the rapid formation of mammals, together with corresponding changes in plant life (Eldredge and Gould 1972).

Natural selection has become a basic concept of modern biology, one that accounts as much for the emergence of every biological novelty as it does for successful adaptation. Survival profiles in Darwinian concepts refer to bodily dispositions modified by contingencies such as the life cycle of a species; in other words, like individuals, species have a developmental sequence in that all species move toward a pattern of extinction. Darwinists treat natural selection with its catch phrase "the survival of the fittest" as being a fully purposeful zero-sum game between individuals in a species in which selection occurs through direct interaction; the maximal fitness of single individuals in a species cumulatively produces over time a physical ability to reproduce at a superior level of fitness, and so both individuals and species survive the vagaries of nature.

Bateson strongly disagrees, as we have seen, with the zero-sum conditions that Darwinian natural selection embraces. He would agree with the Darwinians that adaptation favors the strong and well adapted in local ecosystems. Yet Darwinians could never conceive, as Bateson did, of successful adaptation requiring a capacity for "larger gestalten" as a condition of being well adapted. An organism cannot be autonomous in its environment. Bateson's fitness criterion of larger gestalt indicates the existence of reflexive awareness in a species in relation to its communicative network of interactions with other plant and animal species. The larger gestalt acquired through learning would be advantageous in times of punctuations, or jumps in coevolution leading to species turnover.

Nor can genome and environment be autonomous from each other. Therefore, coding between genome and environment must in some way be complementary "as if the organism inhabited a box into which the incoming information must be filtered." Somatic changes can partly influence pathways of evolution, but instead of coding a direct description of immediate environment, the feedback through which somatic changes occurs acts at a more concrete level than genetic change. Under localized natural selection, some "fitness function" or habit of individuals in a species determines what the next state of the system will be. And sometimes "the acquisition of bad habits, at a social level, surely sets the context for ultimately lethal genetic propensities" (Harries-Jones 1995, 155; MC, 11/11/74, 250–7b).

Yet, Bateson argued, natural selection is not a determining condition in all evolutionary circumstances, nor is natural selection equivalent in scope to a physical law, as many biologists believe. Biological systems are ordered hierarchically (heterarchically), he argued, but homeostatic fitness values in the upper levels of that hierarchy are not simply repetitions at a higher level of abstraction of the same interactions at a lower level. Each level seems to have its own coevolutionary balance.

A fitness function may be viewed as instantiating a homeostatic value ("habit" or set of values preserving survival), but even if the survival values are thus preserved, they are not explicitly coded so that they are isomorphic at all levels of expression. The reasoning of this pertains to the fact that the cybernation of systems, that is the dynamic of feedback patterns, never has exactly the same effects at each level of the biological hierarchy. Bateson took the famous case of Jacob von Uexküll's wood and deer ticks, the latter being unpopular organisms that can carry Lyme's disease.

> A common example first used by Jakob von Uexküll is the tic's [*sic*] sensitivity to butyric acid. Natural selection has selected for butyric acid detection. The purpose of detecting butyric acid is to detect mammals, who secrete butyric acid and whose blood is a food source for tics. However, there is nothing in the system for detecting butyric acid that refers to mammals or to food or to blood. There is only a triggering mechanism which causes tics to release themselves (and drop from a tree, hopefully onto the passing mammal secreting the chemical) in the presence of butyric acid. The purpose of the individual tic's chemical-detection system is represented abstractly in the super-system of the species as a whole. Natural selection "knows" that butyric acid is about access to mammalian blood, but a tic does not "know." "Purpose" arises from or is represented in the system as a whole, from a "simulated" homeostatic value. We say the homeostatic value is "simulated" because it does not maintain itself from feedback at the tic level, in the way that (it might be argued) temperature feeds back to a thermostat. Rather, the high-level purpose of accessing mammalian blood becomes apparent at the species level because something else (detecting butyric acid) is selected for [= apparent] at the individual level. We have, in essence, a hierarchy of purpose, with grammaticalization at the higher levels ensuring that those levels are about something other than what the lower levels are about. . . . In this way, very simple informational decisions at a low level ("drop if you detect butyric acid") can have very large epistemological effects ("know that you need to eat mammalian blood") at a higher level. (MC, 30/05/1967, 1519 e,f)

Bateson's not-so-hidden message is that the course of evolution is much more indeterminate than natural selection suggests because systemic informational effects do not necessarily match informational effects in any one particular level; the whole is not made up of an added concert of parts. The successful phenotype, the one that survives, is a phenotype that exhibits flexibility at several levels, including, of course, that of communication with other members of the species, and interspecies communication. And if timing manifests itself, this means that some higher levels have very slow-moving variables in a hierarchy, while the more immediate levels are faster cybernetic circuits. Then Bateson's arguments begin to make sense of selection as occurring through segmental differences in hierarchical timing. The concept of temporal "fit" is a low-level analogue for "matching flexibility." Bioentropy is thus *a reflexive source* of adaptation for operational conditions of living, and survival depends on those who best manage to retain flexible interactive strategies in the temporal hierarchy of environment.

At the same time, there is a finite amount of *potential* changes that the organism is capable of achieving with adaptive change.

> In biological evolution, adaptive changes occur during the lifetime of an individual [i.e., phenotype], adjusting him or her to various forms of stress, efforts, demands placed upon skill and the like. What is consumed is [bio] entropy i.e. uncommitted possibilities for change in many different physiological and neural variables and parameters. The uncommitted alternatives ([bio]entropy) are lost, eaten up by [genetic] commitment and by becoming unchangeable parts of patterns. . . . Adaptive changes limit the possibility for future adaptation in other directions. (Bateson 1991, 209)

Finally, species can get into trouble when stress creates contradictory demands upon some variable in total physiology. Such a double contradiction—double bind—occurs if the two stresses appear at different levels of overall organization with demands that one variable be increased to meet stress A and simultaneously decreased so that stress B may be reduced. This demand for double oscillation at different levels of organization will be fatal.

VARIETY AS DIFFERENCE

The issue of replacing Darwin's material vision with one centered on appropriate concepts of information and communication had troubled Bateson ever since he had left psychotherapy and gone back to his study of animal communication. Until the period immediately prior to writing "A Re-examination of Bateson's Rule" (2000, 379–395), he was himself not sure how the cybernetic notions of information could be translated to the natural world. He was touching on old dilemmas. The problem facing him was that any explanation that does not rely exclusively on embodiment of physical mechanisms becomes associated with vitalism. An example is that of "entelechy," put forward by Hans Driesch in the nineteenth century, an argument about how nonphysical forms of regulation organize physical processes. Bateson could not find any justification in either Driesch positions, and he wrote that Driesch had put forward the idea that the ultimate pattern determines its antecedents but that "The ultimate pattern never determines its antecedents. There are no final causes." Thus entelechy was a mistaken principle of order.

Nevertheless, it was still possible that patterning, shape, and order flow from apperception of form, and therefore information of systemic importance travels through redundancies in every aspect of the system. His own path of understanding formal embodiment rather than mechanistic embodiment went from forms of learning to behavior, to cybernetics, to morphogenesis and back again. The initial problem he had worked on at the Veterans Administration hospital in Palo Alto was how cybernetic signal events triggered meaning. Information sent and received in classical cybernetics was almost always as a physical signal in a circuit, such as instructions for technical machines, or as "news" in a spoken or visual interaction. In this initial or first-order cybernetics, cybernetic events were enfolded in time series in a channel. In cybernetic understanding, the time series is error-activated, so that corrective action is brought about by the difference between some present state and some preferred state. First-order cybernetics then becomes a model for communication between two people. Later, the technical term "information" became succinctly redefined not only by its feedback properties, but also by Bateson's recognition of the perception of any difference that makes a difference in some prior event: "This definition is fundamental for all analysis of information in cybernetic systems and organization" (2000, 381).

Thus, if information is considered as variety, as in Ashby's definition of information, and if the information event is some type of selection of form in variety, then there is a second-order level of apprehension of messages, and at second-order level, the "form" of the form is "triggered by difference." An information event at the second-order level triggers error correction in a circuit, but any form interacting with any form (not necessarily a linguistic form) generates a pattern of contrast or comparison that embodies a difference. This difference indicates what the signal, or message is "about." It is a sign. Dealing with information at the second-order level, as the form of a form, a sign triggered by difference, yields a much wider range for comparison and contrast than does the idea of information tied in with specific feedback channels. It indicates that information, considered as a sign of "difference," could emerge from simple acts of selection or percept comparison in any context of natural order:

so long as we talk about difference do not pretend that difference is somehow "physical" . . . difference is neither in the outside world, nor solely in the inside world but is created by an act of comparison and this act is an event in time—an act of scanning. Whether there are static differences "out there" it is not so important for us as psychologists as the generalizing that only changes can enter into our perception . . . any difference which makes a difference is "information" (there may be other differences, but these do not concern us as psychologists). (MC, 1971, CAF 308)

In addition:

> Only news of a difference can enter into man's sense organs, his mapping, into his mind. Only difference can effect and trigger an end organ—so all our information (our universe of perception) is built on differences. Difference is "super-natural" i.e. outside the natural world as this is seen by the hard sciences. Difference is not located in x or y or in any space between. (*Notebook*, 51/Fall 1973)

The new way he had devised of dealing with information as the form of a form "triggered" by difference gave an enormous flexibility to his whole construct. Bateson could now argue that "message" or "news" need not be tied to specific circuits of senders and receivers. The triggering of "difference" could occur in any information context through the simple means of comparison and contrast. These could include any means through which a process of perceiving difference occurred in animal interactions or, beyond this, to organisms in the natural world. He could begin to think about contrasts in any form in nature, even how pattern is derived from contrasts and interconnection in morphogenesis. The supposed mystery of patterning of form in nature adapting to its environment could now be seen to be the embodiment of difference in patterns of relationship of organism and its environment; and, in these embodied patterns, living forms created their own organization (Harries-Jones 2010).

His essay "Bateson's Rule," a tribute to one of his father's ideas, showed how he might move from cybernetics to morphogenesis and at the same time pick up some of his father's ideas of the importance of form. Bateson proposed that symmetries and asymmetries of forms can occur as a result of natural forms receiving information from the outside (an unfertilized frog's egg receives information from the entry of the spermatozoon), or it can occur inside the system, as a result of the natural propensity toward redundancy of form in natural systems. Redundancy is the rule rather than the exception in biological systems, and the presence or absence of redundancy in natural form could help explain how symmetry and asymmetry can occur. In other words: "In a plant, the morphology of the fork provides information enabling the flower to be not radially but bilaterally symmetrical, i.e. information which will differentiate the 'dorsal' standard from the ventral lip of the flower" (2000, 395). Complex redundancies thus enable a process that is an analog of ringing the changes throughout the organism, much as the pealing in the bell towers of churches before a Sunday service.

With his concept of difference revamped, Bateson was now in a position to deal with development inside the organism. Even morphogenesis, the growth of an embryo, which many evolutionary biologists supposed to be an entirely material process emanating out of genetic templates, is a developmental process of pattern formation. He argued in one of his later papers that the interaction of units of embryology are not constructed, as in machines, through repetitions of parts joined together to fit as a whole, but could be described better in terms of the contrast of "questions and answers" triggered by difference of comparison and contrast. The unit of embryology is neither just the egg nor just the sperm, each with its individual characteristics considered separately. Evolution has a whole temporal history, so the genetic code in the unfertilized egg has sufficient information to pose a question. It sets the egg to a "readiness to receive" a piece of information. Since the genetic code itself contains no answer to a question, the genetic code "must wait for something outside the egg, a spermatozoon a camel-hair's fiber, to fix it. In order to develop, that unit of morphogenesis is the egg plus the answer. And without the egg plus the answer, you cannot move on to the next phase" (1991a, 179). His essay contains a number of other examples of contrast between "inside" and "outside" as triggers for variety in natural form, and in each case pattern formation in the embryo is presented less as production of organic material and more as the equivalent to the development of an idea.

At the end of his life, he wrote to a distant cousin, the ethologist Patrick Bateson, "Wouldn't it be nice if our colleagues discovered that . . . this system is not a list of organs or indeed any aggregate of material matter—it is a pattern—an idea, [and] the step from morphogenesis to behaviour and learning is a relatively short one" (MC, 04/01/1980, 127–5b).

CREATURA AND PLEROMA

It was a bold move on Bateson's part to draw on a text by Carl Jung in order to present his argument for linking pattern formation to the idea of the nonhuman animal world exhibiting intelligent awareness. The terms he used, "Creatura" and "Pleroma," originally appear in the Jungian text "Septem Sermones ad Mortuos" (Seven Sermons to the Dead; Jung 1965, 378–390). The text itself was one that Jung wrote of his own psychic self-exploration (metanoia) when he was breaking away as a disciple of Sigmund Freud in the years 1913–17. Jung held this text back from publication for many years.

Jung had used "Creatura" to indicate that the essential characteristic of human nature is to distinguish and to make judgments about the basis of human qualities. Jung described "Pleroma as the 'nothingness' or 'fullness' of the eternal and infinite, within which all contrasting qualities are balanced out—good and evil, beauty and ugliness. Pleroma is the realm of homogeneity and the infinite that therefore cannot be distinguished. Creatura, by contrast, is confined to space and time. Creatura creates the qualities of good and evil, of beauty and ugliness, 'in the name and sign of distinctiveness.'"

The most evident contrastive aspect is that Creatura is a figure that "speaks of itself." Jung wrote: "Distinctiveness is creatura. It is distinct. Distinctiveness is its essence, and therefore it distinguisheth. Therefore man discriminateth because his nature is distinctiveness. Whereof also he distinguisheth qualities of the pleroma which are not. He distinguisheth them out of his own nature . . . speaking from the ground of our own distinctiveness and concerning our own distinctiveness" (Jung 1965, 380). Bateson took this to mean that Pleroma is background from which Creatura emerges. Pleroma is "the unliving world described by physics which in itself contains and makes no distinctions." It follows from this that if Pleroma is "void" or "fullness," a totally unstructured realm about which nothing can be said or thought, we can only know the nonliving material universe indirectly, through a methodological framework we ourselves establish. The description of Pleroma undertaken by physicists, for example, explains "appearances" of Pleroma through the dimensions of mass, time, and length, but it remains an "as if" world and not a world of difference: "I can describe a stone, but it can describe nothing. I can use the stone as a signal—perhaps as a landmark. But it is not the landmark. . . . What happens to the stone and what it does when nobody is around is not part of the mental process of any living thing. For that it must somehow make and receive news" (1987, 17).

Creatura, on the other hand, is "that world of explanation in which the very phenomena to be described are among themselves governed and determined by difference, distinction and information." To say anything is at once to create distinctions and is therefore the ground out of which creatura looms. Creatura is a figure in the ground of pleroma—not a dual or nor even a polarity, but contrastive, as in gestalt. In distinguishing, Creatura distinguishes through the distinctions of relations that it draws toward, and so recursively points to the criteria for distinguishing. The configurations of Creatura are pointers to how we see differences. Thus "creatural theory" becomes an epistemology contributing recursively to "the epistemology of how we know and think." (MC, 1975, CAF 69)

The use of these obscure terms risked the charge from fellow biologists that Bateson was presenting a vitalist argument. And when any biological argument is deemed to be vitalist, biologists begin "behaving badly," says Susan Oyama, for such a suspicion, true or not, contaminates by innuendo (Oyama 2010, 401). Biologists have been quick to label many alternatives to biological mechanism with the term "vitalism." To choose such terms as Creatura and

Pleroma raises the specter of scientific heresy, of an immaterial principle, a life force, that inhabits and animates the body, and provokes the notion that Bateson is a suspect biologist talking about a mysterious essence. Even if few biologists today share the premises of Driesch metaphysical vitalism, such terms, along with "holism" and "organicism," are all suspect.

At the same time, Oyama notes, physicalist biology has given itself permission to reconfigure terminology at will when using mechanistic explanation. Molecular biology has taken over terms such as "translation," "transcription," and "editing," in order to explain activity in the genome. It has done the same with "sense," "nonsense," and "reading frames." So what does an inquisitive biologist do, asks Oyama, when all the good words are taken into a technical vocabulary and given explicit definition? While the same terms continue to be used by ordinary people in relation to "meaning," molecular biology insists all the while that its own borrowed terminology is technical and exists as technical phrases that have nothing to do with meaning.

Biologists are, generally speaking, ontologists, which perhaps accounts for their wish for clearly distinguishable types between mechanism and nonmechanism. Bateson, on the other hand, was drawing on epistemological difference, in order to discuss how a scientist might go about choosing methods to explain the presence of life. His epistemological framing for this task becomes clear and specific when he outlines his six criteria of mind: (1) A mind is an aggregate of interacting parts or components; thus, Creatura always exhibits a sequence of interactions between parts. (2) The interaction of parts of mind is triggered by a perceived difference. (3) Mental process always requires collateral energy. (4) Mental process requires circular (or more complex chains) of determination. (5) In mental process, the effects of difference are to be regarded as transforms (coded versions) of events that preceded them. (6) The description and classification of these processes of transformation disclose a hierarchy of logical types immanent in the phenomena (Bateson 2002, 92).

It is for epistemological reasons that Bateson includes logical typing in his criteria of mind (criterion 6). The description and classification of these processes of mental transformation disclose a hierarchy (heterarchy) of logical types immanent in the phenomena in which: "metarelations between particular agents may be confused but understanding may emerge again as true at the next more abstract level" (Bateson 2002, 118). In these cases understanding begins in hypotheses, which are then checked or corrected by sequences of contexts of meaning. The context itself refers to prior interactions and recalls the contexts in which they took place so that discrimination can distinguish "this" from "that" according to difference in the formal aspects of each previous context. Criterion 6 ensures that Creatura and Pleroma do not come too close to being ontological distinctions. Ultimately they are metaphors that encapsulate distinctions in a gestalt embodying mind and matter.

Oyama, a developmental biologist, believes that her own branch of biology, to which Bateson himself was attached, tries to do justice to the astonishing alterations in time and space that constitute developmental systems, without resorting to any notion of "soul" or a "Maker," or divine plan. The focus of developmental systems theory, she says, is on transactions across boundaries at many scales, on relations between things and events, and not spontaneous emergence "within an essence," which was the hallmark of vitalism. Yet modern biology is exceedingly reluctant to permit any excursion into metaphor because it feels that the history of vitalism has cast a shadow on its strict approach in the use of scientific method. The result, in Oyama's view, is that many biologists hold fast to a view of "biology [which] is constituted *for* and by a mechanical, mechanistic model which is itself blind to the concept of life" (Oyama 2010, 418).

If developmental systems theory succeeds in explaining interactions in developmental systems, the approach and its theory would demonstrate how organisms in transition are the product of the processes of "nurture" (Oyama 2010, 407). The term "nature" would indicate processes complementary to this outcome and would share the same emphasis on temporality and transformation, as does morphogenesis. Discussion of "nature" in developmental biology must not shrink away from the possibility of being called vitalist because of its focus, and it must continue to pay close attention to the many alternatives to mechanistic biology, of which Bateson's set of ideas is one.

The following two chapters will go outside of Bateson's writing to consider other authors' contributions, like those of Oyama, to Bateson's ideas. Some of these authors have created their themes independently of Bateson, while others, like Oyama, acknowledge his contribution in relating pattern to process. An example is that of the timing characteristics of morphogenesis, an issue that Bateson mentioned but was unable to give any extensive examination. As a developmental biologist, one could make the case that timing is important enough to be considered as the seventh criterion of "mind." Timing cycles concern duration and change: the combination of life and death whose oscillations determine the time grains, rhythms, and size grains of biological systems.

The reason for Bateson's interest in timing cycles flowed from his own increasing dissatisfaction with using Bertrand Russell as a mathematical model to support his notion of logical types. "Double bind theory must be detached from the Russellian logical types and realigned to derive perhaps from the theory of mathematical groups," he wrote to the British anthropologist Edmund Leach in 1973, noting that this would be "the focus of my thinking and teaching in the remaining years" (MC, 15/03/1973, 824–2). He declares that Russell had been "wrong in the sense that the world in which we live is not organized in lineal, transitive relations" and "if the chains of causation are to be plotted on to something like chains of logic then the model must be a recursive one" (MC, The Evolutionary Idea Glossary, Box 5). Presumably this acknowledgment of error was to some degree a self-acknowledgment of his own error in sticking to Russell's explanatory scheme for too long.

In mitigation of his error, it is evident that for many years he was hoping to secure "a body of theorems analogous to the calculus of propositions i.e. to perform for cybernetic causal systems (which contain time) what the calculus of propositions does for logical systems (which are timeless)" (MC, 21/02/68, 672–17). The receiver of this letter, Tolly Holt, was working on Petri nets, which are indeed a means of incorporating temporality into the structural ordering of living systems, but Bateson felt unable to take on Holt's work. He turned therefore to a different direction, toward mathematical group theory, which might help him link in a more precise manner interrelations between subgroups to the interrelations of larger groups of which they were members. Group theory might explain more adequately how parts relate to wholes in ecology and what sort of transforms are possible in the expansion of any subgroup to a larger group, and in this process, the sort of limitations that might apply in such a transformation (MC, 1956, CAF 208).

The relation of parts to wholes becomes even more pressing once Bateson moves into the study of ecological order. Once he abandons Russell's lineal analysis, Russell's series of transitive steps in hierarchical order becomes replaced with recursive loops, loops that occur in timing cycles but that are relatively invariant in ecological order. This presupposes a topological logic of connectivity similar to that which orders morphogenesis, in that any movement of the part is in some manner preconditioned by movement in the whole.

> What I'm saying is that the world into which we are moving, the world in whose terms we have to think, is a world of patterns and in that world are tautologies and logics which we can use for explaining, for building accurate language and for creating some rigor. It's not like the language of quantities and such things. It's a language of patterns, and, for most of us an unfamiliar business. (Bateson 1991, 180, 184)

7

A POSTGENOMIC VIEW

> Before Lamarck, the organic world, the living world, was believed to be hierarchic in structure, with Mind at the top. The chain, or ladder, went down through the angels, through men, through the apes, down to the infusoria or protozoa, and below to the plants and stones. What Lamarck did was to turn that chain upside down. He observed that animals changed under environmental pressure. He was incorrect of course, in believing those changes were inherited. . . . When he turned the ladder upside down, what had been the explanation, namely the Mind at the top, now became that which had to be explained.
>
> —GREGORY BATESON

Problems always arise when analyses of traits and behaviors of individual organisms at a single level are used to explain very broad systems. Biological systems do not connect their parts in any one-to-one, point-to-point manner, as is typical of mechanical systems. Nor is it possible to explain evolution as an outcome of single events, steered only by natural selection. Bateson asked his readers to look at evolutionary processes as a series of coupled events—more specifically, coupled communicative events. For example, a Darwinian interpretation of the evolutionary adaptation of "horses" and "grass" explains the adaptation of "horse" and that of "grass" as being the adaptation of separate individual things, and fails to evoke the feedback relation between them. It fails to understand that grass responds to horses" hooves as much as horses respond to sweeter grass, instead "the horse" and "the grass" are presented as duals involved in a competitive relationship between "horses" and "grass" (MC, 04/03/1972, 1209–11).

Bateson offers a definition of coevolution as "a stochastic system of evolutionary change in which two or more species interact in such a way that changes in species A set the stage for the natural selection of changes in species B. Later, changes in species B, set the stage for the selecting of more similar changes in species A" (2002, 227). In other words, all changes in coevolutionary settings are really moves in the relationship of organism and environment to preserve a relationship, and to stabilize that relationship through adapting to continual variance within it. Evolutionary adaptation is always of organism plus environment.

Coevolutionary change depends upon two contrasting sets of processes, with the one set conserving developmental regularities, and the other facing outward toward the vagaries and demands of the environment. The two sets of processes in a coevolutionary framework contrast with each other both in time sequences, such as adapting to fast and slow variables, and in the way in which processes of selection work. Bateson incorporates these notions into his discussion of his six criteria of "mind" (2002, 92).

Some biologists have taken Bateson's coevolution to mean that he is supporting Jean-Baptiste Lamarck against Darwin, by supporting Lamarck's hypothesis that experience, performance and use of bodily parts creates conditions for evolutionary change. Lamarck had cited a giraffe developing its long neck through continual stretching in order to gain access to higher stands of leaves on acacia trees. He argues that the stretch in the giraffe's neck is an example of how variation develops among living organisms through the ability of parents to endow their immediate offspring with advantages acquired in parental behavior. The Darwinians took this as being a clear example of Lamarck's not knowing anything about genetics and therefore not knowing what he was talking about. They were correct. They also accepted the point that Lamarck, too, thought of the giraffe's neck and the leaves of the trees as separable things. But Bateson argued that the real point worth noting was that Lamarck recognized that giraffes, like other organisms, have perception and—through their perception—have developed an apperception of their environment, or minimal intelligence.

Bateson himself had no interest in Lamarck's argument that the flexibility gained through an organism's experience is transmitted from parents to their young within a single generation. In fact, he offered a counterargument to this aspect of the Lamarckist position, arguing that if an inappropriate response in a stressed situation is inherited in the way that Lamarck proposed, then over the long term it may add to, rather than correct, acquired unfavorable conditions encountered in the parental generation. The giraffe's body, like the human body, is made up of a very large number of variables, which interlock in all sorts of spirals and loops. If error correction consisted of accepting continual short-term adjustments, all Lamarckian heredity might do is to enforce an ever-increasing rigidity. Once Lamarck's organisms started locking themselves into any one circuit of repetitive activity, through genetic inheritance, individual giraffes would end up "stressed out," with no tolerance or flexibility anywhere. The generation that had accepted one adaptation in their circuit of repetitive activity, which they then would pass on to the next generation, would increase the chances that one spiral of tightening will ipso facto tighten the others, so that eventually such a sequence of adjustments would lock or restrict total system performance. The eventual result would be a total loss of ability to adapt.

An example that Bateson sometimes used was that of Lewis Carroll's "Bread and Butterfly." Lewis Carroll's version of Bread-and-Butterfly was written in order to present an ironic depiction of the suspect logic in Darwin's natural selection, in which it would seem that fitness is acquired piece by piece over time as a series of metaphorical bandages. Carroll's humorous portrayal of Darwin applied to Lamarck as well, said Bateson. Bateson's own criterion of flexibility requires multivalent response and not repetitive mitigation through one-by-one somatic or genetic changes:

> . . . we still have time to consider a fictitious animal which has long fascinated me and is relevant to this whole business of adaptation, addiction, and double binds as a possible source of positive advance [in our understanding]. I refer to Lewis Carroll's Bread-and-butter-fly [and Carroll's text reads]:
> "Crawling at your feet," said the Gnat (Alice drew her feet back in some alarm), "you may observe a Bread-and-butter-fly. Its wings are thin slices of bread-and-butter, its body is a crust, and its head is a lump of sugar."
> "And what does it live on?"
> "Weak tea with cream in it."
> A new difficulty came into Alice's head. "Supposing it couldn't find any?" she suggested.
> "Then it would die, of course."
> "But that must happen very often," Alice remarked thoughtfully.
> "It always happens," said the Gnat.
> After this, Alice was silent for a minute or two, pondering.
> (Carroll, *Through the Looking-Glass*, quoted in Bateson 1991, 211)

Bateson's own depiction of natural selection was that of "picking out the unfit for non-survival" (MC, 11/06/64, 463–4b), where the "picking out" refers to a failure to maintain stable relations between the organism and environment while both are undergoing change. When we ask what caused the Bread-and-Butterfly to die in Bateson's playful portrayal, we have to answer that the fly died not simply of the peculiar trauma of having his head dissolved in weak tea, nor yet of simple starvation, but of the seeming impossibility of meeting the changing circumstances in which a required double adaptation could take place. A situation had emerged where if one failing did not get you, the other one, at a higher level, would. Dilemmas of adaptation therefore, as Bateson pointed out, are very close to the dilemmas of double bind.

Alternatively, when each coadaptation is defined as being like the changes in "horses" and "grass" relationship, each change indicates moves necessary to preserve the continuity of their relationship with each other, both horses and grass in the wider condition of a relationship in changed environmental surround. Our own percepts of nature should preserve this understanding of the coevolutionary continuity of relations within nature and its systemic characteristics. This might lead perhaps to a redefined notion of "natural selection." There is a warning attached to this proposal: inappropriate understanding of coevolutionary relations compounds errors. The compounding of errors of perception of coevolutionary change will lead to quicker destruction of environment than if the Darwinian argument had been correct.

According to an unpublished manuscript of his entitled "The Evolutionary Idea," Bateson's discussion of coevolution was to have included a lengthy commentary on "appropriate syntax." As we have seen, the "syntax" of the dominant view in biology weds its conception of embodied energy in the environment to a "field" of complex physical resources, biomass, and bioenergy, whose operations functionally interact to provide physical necessities of life. On the other hand, which difference makes a difference in any ecosystem is "a threesome business," he wrote, and is a relational triad. The first term of the triad is circular causation at several levels, which must therefore include morphogenesis, the generation of form per se. The second term is that of co-learning in adaptation, which would include co-learning between animals and their environment. The third term is that of genetics, which is a memory system, and a much slower temporal variable than the other two. Each draws upon the other recursively as relations of the other. Bateson goes on to argue that the study of coevolution is a means for unearthing all the processes we call "knowing," and that evolutionary theory, will consist of subject-predicate sentences in which the subject will always be a relationship and not an object (Bk. Mss., Box 5, 1987, 205.27).

COEVOLUTION AND MORPHOGENESIS

The term "mutualism" seems to be more acceptable in biology than "coevolution," specifically in the literature on biodiversity. One might suspect that the terminological preference is associated with the way in which mutualism can be associated with "productive efficiencies." Mutualism binds together scattered species so that they create the basis for "an efficient ecosystem"; the idea that neither can independently exist seems secondary. For Bateson change in coevolution did not require such massive ongoing change in productive efficiencies, for the triggering of "difference" could occur in any information context through the simple means of comparison and contrast. These could include any means through which a process of perceiving difference occurs in animal interactions.

He could now begin to think about contrasts in any form in nature, even how pattern is derived from contrasts and interconnection. The supposed mystery of patterning of form in nature adapting to its environment could now be seen to be the embodiment of difference in patterns of relationship between the organism and its environment; and through these embodied patterns, living forms create their own organization. In a coevolutionary setting, such changes could be initiated either from the inside or from the outside. His essay on "Bateson's Rule" shows that symmetries and asymmetries of forms can occur as a result of natural forms receiving information from the outside (an unfertilized frog's egg receives information from the entry of the spermatozoon), or they can occur inside the system, as a result of the natural propensity toward redundancy of form in natural systems.

Redundancy is the rule rather than the exception in biological systems, and the presence or absence of redundancy in natural form could help explain how symmetry and asymmetry can occur. On several occasions, Bateson points to the study of morphogenesis as an exemplar. Morphogenesis is the biological process through which an organism develops its shape, and is coincident with ontogeny, the development of an organism from the fertilized egg to mature form. The process is a prime example showing how communication procedures are apparent in biological contexts as analogies of human question-and-answer communication.

Morphogenesis, like other developmental processes, has its own general rules and properties, which bump against molecular biology's genetic determinism. Systemically, morphogenesis is a system of nested sets and part-wholes (Salthe 1985). Unlike molecular biologists, those interested in morphogenesis pay particular attention to finding patterns where different rules and concepts operate among many levels. Developmental biologists, for instance, accept that genetic activity applies to the special circumstances of the genomic level, but hold that developmental processes from thereon, after conception, involve many other levels of organization besides that of genetic activity.

Gene processes are carried forward to the organism level. Here, new developmental processes, based on gene-level patterns, build new organism-level patterns and these then become the raw material for yet higher level processes acting at the next level. A "level" could well apply to activity in the embryonic stage inside an organism, or to genetic variation expressed at a more general level through genetic populations rather than genetic individuals. But a level may also apply to phenotypic variation at the individual level, or landscape variation only at the landscape level. The convention within morphogenesis is that a morphogenetic system is one with many levels, yet each level can have their own structural properties. Each will reflect processes acting on preceding levels so that a species level is not too disparate from the level in which phenotypes act as "individuals" (Thompson 1988, 14, 18). As Bateson pointed out, part-whole relationships are not simply that of parts put together to function as a whole, as in mechanical devices, but are hologrammic in the sense that the pattern of the whole enters into the parts as well as parts enjoining themselves into a whole unit as well. In addition, processual activity in morphogenesis is not fully determinative, but proceeds in the shape of a "reverse hierarchy" that is with outlines first of all, before it "fills in." The shape of morphogenesis, in its genesis, follows a form reminiscent of the formation of a perceptual gestalt.

On this aspect of patterning in morphogenesis, the developmental biologist K. S. Thompson gives much fuller detail than does Bateson, and in doing so gives depth to Bateson's arguments. For Thompson, pattern is represented by "ontogeny," and a pattern of "form changes" or "form stabilizes" can only be recognized over time. Control and constraint in pattern formation result from the establishment of, and response to, *whole organism properties*, such as fields and gradients, and then from more localized patterns and reference points. In much of this, the oocyte, or fertilized egg, and the cells of the early embryo act as if they have "knowledge" of their positions in time and space—or what Bateson referred to as a "readiness to respond to information." Unless the respondee already has a certain degree of contextual understanding, response is unlikely to occur. Positional information represents a response to pattern-generating and controlling mechanisms operating in the embryo, from egg to late morphogenetic stages (Thompson 1988, 35).

There is a distinction between pattern and process. Ontogeny is not a "mechanism" of development; therefore it cannot be translated into concepts of mechanisms of evolution. It is a pattern that can only be compared with other concepts of pattern, such as "form changes" or "form stabilization" (phylogeny), over evolutionary time. Process, on the other hand, involves evolutionary interaction generated through various types of feedback, including those of pattern. Bateson's particular contribution to the study of morphogenesis and to developmental biology as a whole, Thompson states, is the recognition that any interactive system with incipient hierarchy of levels has contrasts between actual patterning and its processual operations. Its processes are so complex that they can only be resolved, or conceptually understood by an observer, when that observer recognizes that the interactive system of ontogeny is self-organizing (Thompson 1988, 19).

Interdependence, interaction, and contrast in pattern and process occur in the form of feedback among the separate levels of biological systems. This is why no biological system replicates the top-down hierarchy of political or military systems of command and control. At every moment in a developmental system, new information is being added in the form of new gene expression. This is probably the only way that a truly complex organism can be created—through an exponential geometry of information—that is, through complexity growing out of itself. Information systems are unlike mechanical systems. Complexity in mechanical systems occurs through the addition of subcomponents, which then have to be integrated into the whole system. Information systems by contrast are multiplicative, not additive; changes ramify through the whole system, including feedback from the whole system unit. Thompson notes:

> The amount of information in such complex and interactive self-assembling and self-regulating systems is far greater than the arithmetic sum of discrete bits of input. Even so, many phenotypic features are not fully determined at the genetic level. A classic case is that of fingerprints . . . the amount of discrete genetic information that would be needed to code for such a unique phenotype for each individual is impossibly large . . . fingerprints self-assemble during development . . . the same is true of the detailed "wiring" of the human brain. (1988, 71)

Definitive instructions prescribing the fate of given cells and their descendent clones are given late rather than early in the sequence of development. Instead of the fate of every part of the embryo being irreversibly determined in advance, they are actually acquired as development proceeds, with fingerprints being one of the features that become laid down last of all. Indeterminate embryos are not only self-regulating but, in every sense, self-assembling. Yet within multilevel biological systems, the interrelationship of pattern and process—and their contrasts—carry forward to the next level so that control over the processes of development is created progressively, rather than being predetermined in the oocyte and fertilized egg.

Each increase in information is immediately multiplied many fold at the *epigenetic* level as well, because of the way in which the new information changes the nexus of interacting signals. In fact, there must be "simultaneous operation of processes at several levels." In addition, the results of processes at each level directly influence processes at other levels; they contribute to "causation" at other levels. Processes occurring at one level constrain other levels, either through upward "causation" (i.e., consistencies in pattern of feedback) or through downward causation, affecting levels below. Upward or downward causation may be due to either the process of introduction of variation, or its sorting, or both.

This ongoing recursive feedback between the genetic and epigenetic components of the developing system is central to the whole system undergoing development. Information is added to the system in the form of expression of particular gene arrays. These then modify the immediate epigenetic environment. The environment then calls forth a new set of gene expressions and so on. As the embryo moves in its steady progress toward the final determination of all cell and tissue types, each new level requires irreversible "decisions," for such a system operates through a series of sequences in levels, each dependent on the one before:

> The simplest and most useful way to envisage the control of development processes is as a cascade of "decision points.". . . At each decision point each cell and lineage encounter a "switch" and respond to one of the two or perhaps even more possibilities the results of which (a) initiate new gene expression or (b) materially alter . . . the behavior of the cell during morphogenesis and (c) progressively restrict the potential range of future differentiation of that cell and its lineal descendants. [Toward the end of this long chain of events, cell lineages actually become determined.] The accumulation of genetic expression via the passing of these switch points constitutes a "memory" (i.e., positional information or genetic address). (Thompson 1988, 70–72)

Yet sequences and activities in these developmental pathways are subject to constraints, which Bateson sometimes refers to as "rules." As in all complex systems, the combination of gene expression and sequences in developmental pathways is affected by initial conditions. Slight variance here can result in different outcomes. Initial conditions are, as it were, "the rules of the game." As Bateson pointed out, if we are talking about mind and information we are talking of a logical product that generates a difference, and a logical product is not additive in the simple sense (1991a, 163). Then pattern in the embryo will subsequently be constrained by the developmental rules appropriate to the given focal level of the newly defined cell lineages. These constraints may be strictly defined and mathematically evident to an observer as in the case of the patterning of leaf. In leaf patterns, growing shoots or flowers follow the classic mathematical rules of the Fibonacci series, that is to say they exhibit fractal dimensions. The fractal dimensions appear to be primordial events.

Developmental events will also be constrained by the boundary conditions within which these "rules" operate for the given set of cell populations. Yet developmental rules do not result in deterministic outcomes since the combination of boundary conditions, initial conditions and the conditions in which the events themselves occur—the "players in the play of the game"—are unique to that particular event. In addition, there is another type of constraint in developmental pathways, resulting from feedback in new gene expression. Timing is of particular interest here because it links developmental processes to evolution in an understandable way. Technically, rules of cell-to-cell contact, relationship of cells to substrates and to their interactions are among the most important rules. But other rules of gradients, of waves and of timing, produce higher-level rules of pattern control and timing. All affect relative growth and allometry.

By no means do all biologists who study morphogenesis subscribe to morphogenetic development as the development of communicative forms. Nevertheless, they are more ready to subscribe to the notion that morphogenesis demonstrates the whole entering into the parts of the whole of the developmental process, which is certainly why Bateson chose morphogenesis as his example. In other words, the pattern of events in morphogenesis, as with co-evolution and with all conditions of learning, is of a pattern of iterative events, or recursion. Recursion is basically a turning, a "rotation" in loops of feedback, and is characteristic of dynamic processes of complex systems through which complexity grows out of itself. Recursions are probably the only way a truly complex (organic system) can be created—through an exponential geometry of information. What sort of processes of recursion may, therefore, be regarded as its typical features, and in what way does discussion of the recipe for embryonic development, the embryo interpreting the environment of that which it is part, differ from other rules of biology? The fundamental answer seems to be that genes do not control; they cooperate in producing variations on generic themes produced by the dynamics of morphogenetic fields (Goodwin 1994, 41). And this characteristic finds concurrence among a wide section of developmental biologists, including Bateson, Oyama, Thompson, and Goodwin, as we shall now find out.

LAYERS AND RHYTHMS

Bateson was not the only biologist to be labeled "Lamarckist." The neo-Darwinians took rhetorical advantage of their developmental biology colleagues during the Cold War with the USSR and did so to advance their own emphasis on genetics as the determinate aspect of evolutionary change. In the process they demonized Lamarckian evolution. They made the cooperative and experiential aspects of Lamarckism, implying collective utility, the scarecrow of evolutionary theory and alleged that Lamarckist-type developmentalism was politically linked to Soviet biology. This politically contrived attack resulted in mid–twentieth century formulations of nature vs. nurture in its crudest forms (Fox Keller 2006, 291).

Developmental systems theory (DST), with its emphasis on morphogenesis, was one of the most prominent groups in biology that had to suffer this propaganda. It was the position of DST that any theory of morphological evolution is incomplete if based on Darwinian mechanisms alone. "The hegemonic notion of the genetic program [in molecular biology] essentially foreclosed a way of conceiving morphological continuity and change across generational line in anything but genetic terms, and [yet] there are no consistent rules for mapping genetic mutations to morphological change" (Newman and Müller 2006, 41).

> Against the molecular biologists those in DST agreed that while information required to make a complete organism is evidently contained within the genes of the genome, the genes alone are functionless. Genes require a complement for their performance. Without the cell, the genome is nothing, for the cell provides the context for the activity of the genes. The genes "need . . . transcription and translation that is itself encoded in the genome. Thus a genome can function only in the context of a living cell, which already has the necessary molecules. . . ." (Bürglin 2006, 34)

There are at least two domains that give rise to "contexts" for genetic performance; the first of these is the cell within which the genome lies, and the second is the changing conditions of environment, which is also an active boundary for overall genetic performance. This was essentially Bateson's point: environment and genetics are loosely coupled, both produce stresses on the phenotype, but with corresponding uneven temporal delays. Through identifying genetic continuity as the sole fundament of evolution, the neo-Darwinian synthesis had contributed powerfully to the polarization of debates over the relative forces of genes and environment. Neo-Darwinians had promoted their claims in highly emotional terms, spreading the idea that "heritability" of intelligence, or specific behavioral attributes, is determinately attributed to genetic composition. After the failure of social practices like eugenics, they sought similar determinisms through promoting the idea of a complete reading of the human genome as "The Book of Life," and then to develop social practices around genetic reading of its revelations.

A number of events subsequent to Bateson's death revived the salience of DST positions. Among the first was the widespread interest in retroviruses and their relationship both to cancer and to AIDS. The central dogma of molecular biology did not support the notion of retroviruses. At the turn of the millennium, a presumption of genetics was still that of Francis Crick's "central dogma" of molecular biology, namely a one-to-one correspondence between genes, proteins, and a given trait. The central dogma also held to the idea that once "information" has passed into protein it cannot get out again, so that reverse transfer of specific information from protein to nucleic acid, that is, from environment to genetic performance, is impossible.

The Human Genome Project was solidly behind such premises of the central dogma, yet its results showed that the central dogma had to be revised. Its results showed that a simple mustard weed has 26,000 genes, approximately the same as that of a human being. Clearly processes other than straightforward one-to-one genetic transcription of gene to protein, as the central dogma maintained, were involved in the "Book of Life": "If the HGP is judged by the explicit promises that its proponents made in the late 1980s and 1990s to secure public support (and funding), it has been an unmitigated failure, the most colossal misuse ever of scarce resources for biological research . . . [for] sequence gazing alone cannot predict with confidence the precise function of the multitude of coding regions in even a simple genome" (Sakar 2006, 79, 87).

The failure of the Human Genome Project to substantiate genetic determinism of this type was, to say the least, a shock to mainstream biology. As for some DST theorists who were already disabused of genetic determinism: "The confusion we have experienced as to the nature and the adequacy of the current view of organisms . . . finally led us to the conclusion that the only way of achieving clarification was to abandon the system of concepts which we call the evolutionary paradigm and attempt to construct what seems to use a more satisfactory structure . . . morphogenesis" (Webster and Goodwin 2006, 99). The rehabilitation of developmental systems biology, with its systemic approach to biological processes, began to foster an interest in the sort of questions about pattern, rhythm and variation at a very deep level within biological organisms, an area of investigation that had so fascinated Bateson's father, W. Bateson. As mentioned in the Introduction, W. Bateson had been intellectually "bashed" and "buried" under the early waves of support for molecular biology and genetic determinism. It also raised the salience once again of issues that Gregory Bateson had himself tried to pursue. Among these was field theory. As already pointed out, Kurt Lewin's exploration of field theory in relation to social phenomena had greatly influenced Bateson. Now it was apparent that there were generative fields right at the base of biological organization that have autonomy of movement and are able to create their own patterns out of noise. In the new terminology of DST, life at its base was that of "chaotic patterns."

These generative fields yielded quite a different understanding of how parts related to wholes. Since movement in the whole field influenced aspects of parts of the field, dynamic interrelations became the center of investigation, and the DST field perspective revived the merits of holistic dynamics. While it is true that a change in one gene can make a big difference to the shape of an organism, DST theorists believe that the real role of genes is to select or to stabilize one of the several alternative forms available to the organism as it is developing. The major impact of the gene is always upon the initial conditions of this selective process. Meanwhile, DST pointed out, the large-scale differences of form between types of organisms that can be found in all biological classification systems, require—other than that of natural selection operating on small variations—another understanding of creative novelty in evolution.

DST also believes that if the complexity of evolution within the organism is to be understood, that the DNA of the gene has to be considered within the context of the cell. This is another issue that Bateson's father had raised at the turn of the twentieth century. Examining the DNA in the gene is insufficient to explain all the differential processes that lead to complexity of form in an organism, namely how a heart, a nervous system, a limb, or any other organ of the body develops. While DNA is the only macromolecular constituent of a cell that is accurately copied and partitioned to the two daughter cells during cell division, and is therefore a stable transmitter of hereditary information, DNA is not an independent replicator. It is the cell that reproduces, and in that reproduction it invokes copying and partitioning of DNA.

If DST is correct, then this clearly indicates the need for a layered approach to the study of organization within an organism. Like the DNA, the cell is not the sole contributor to complex differentiation. It too develops within a context, and the context within a metacontext and so forth, such that in the end the whole system evolves as a reproducing unit. Gregory Bateson's point about seemingly infinite nesting of contexts would seem to be justified. While the DNA influences particularities, particularities occur within an environment—and not just any environment, but of a field that is organized in space that has its own constraints affecting the developing embryo. At this point, it is best to stop thinking only about component parts and more about the dynamic organization of the whole. When doing so, the first evident task is to characterize the nature of the dynamics ordering relations within the whole field.

Goodwin demonstrates that the relational "parts" all show similar types of dynamic activity, such as rhythms and waves, that often propagate in concentric circles or spirals. In mammals, for example, this is a characteristic of the activity of a heart muscle, or of neurological activity. There is no overall distinction between mind and body, germ plasm or somatoplasm so far as type of dynamic activity is concerned. Each aspect in the field is one of "excitable media." As excitable media, they all have characteristic patterns:

> Instead of waves propagating from centers, the whole system can change periodically from one state to another so that [its] changes occur in time, but they are spatially homogenous. . . . For instance a wave can be started at one point by an external stimulus, and this can propagate over the whole system as a single widening circle that stops at the boundaries, after which the system goes quiet unless stimulated again. (Goodwin 1994, 54)

If one adds a third dimension to the whole activity of propagation, it is possible to see propagating waves that are like sheets wrapped in a scroll, or even propagating waves that are "twisted scroll waves," in which the generator at the center is not a wave travelling in a circle, but is travelling around as if on the surface of a central doughnut or torus. In any event, there is a change in the pattern of activity from "noise" to bursts of oscillatory activity and the amplitudes of these oscillations vary in a systematic and repeatable pattern. Such patterning is widespread.

The very first descriptions of these "excitable media" were made initially of a set of events occurring in a Petri dish. Later, Goodwin found that the phenomenon is by no means confined to these limited conditions, and can be seen in a variety of evident external behaviors, for example, in the flocking of birds. It has also been observed in colonies of ants where there are sudden transitions to dynamic order affecting the colony as a whole. Here the pattern of activity moves from chaotic to rhythmic and back to chaotic again, repeating this transition at regular periods of time. It has even been uncovered in neural networks during sequences of "learning," suggesting that there are simple dynamic rules covering the sudden transitions. Yet, if there are simple dynamic rules, these cannot be predicted from component parts. Only when both the whole and the parts are studied together and their relational order identified, is there any hope of mapping sudden transitions from such rhythmic activity, and the emergence of subsequent pattern or form.

For Goodwin and others in DST, the real importance of the uncovering of rhythmic phenomena is the realization that chaos or "noise" on the one hand, and pattern or ordered conditions on the other, together with sudden transitional movement between the two, is a phenomenon that is endogenous to life. It is as ubiquitous as feedback or feedforward in Creatura (to use Bateson's term from here onward). Sometimes it is the combination of feedback and feedforward in cytoplasm that generates the conditions for an excitable medium to undertake these unexpected transitions. The sudden transitions both break and restore symmetries in the cytoplasm. And such breaking and restoration of symmetry is crucial to developmental sequences in morphogenesis (Goodwin 1994, 95). Goodwin identifies this as "the moving boundary problem." A developing organism has to generate its own form from a simple, symmetrical initial shape. In order to do so, it has to undertake a transition from a state of higher symmetry (lower complexity) to one of lower symmetry (higher complexity). The technical term for this is to undergo a bifurcation.

One of the means for accomplishing this double transition is through creating whorls:

> The process is like a traveling wave that rises and falls with an irregular periodicity, leaving structural record of its passage in the sequence of its whorls. . . . This type of process is of particular importance in morphogenesis, since the organism changes its shape as it develops. The result is that there is an intimate connection between the dynamics and the form; the field dynamics generates a pattern that leads to a particular shape that then affects the dynamics, resulting in an unfolding of form through a sequence of changes. (Goodwin 1994, 104)

Most important to this postgenomic view is the suggestion by Goodwin that the rhythmic patterns of transition in morphogenesis are a means through which individual organisms sense and "learn." A moving boundary problem, Goodwin notes, is one in which the boundary of the field moves as a result of patterned growth, in this case a growing cell. The whorled theme appears not once but influences bifurcation at each lateral as it moves to smaller elements, and as it does so, gives rise to fractal patterning. In other words, it creates its own structure of self-similarity on successively smaller scales. The cascade of symmetry-breaking bifurcations that forms the basic pattern generator as a result of excitable dynamics, together with a change of shape, result in a morphological theme that is generated through its moving boundary. The changing shape or form feeds back into the dynamics by creating a reference point that stabilizes modes that generate the form, and creating the conditions for the next bifurcation (Goodwin 1994, 107).

Thus the process reveals an order in the dynamics that is then explicated in the form self-similar shapes of the fractal patterning, which in turn then influences continuous replication of ongoing sequences. The generic forms that arise from the breaking cascades of morphogenesis result in species that are stable in some habitats. Yet, this does not exclude a whole new environment of possibilities, as more symmetries become broken, resulting in more complex forms, especially in relation to sensory perception and locomotion (Goodwin 1994, 107, 158).

To say that organisms take discriminating actions that matter to them in their development and for their survival is common sense, says Goodwin. It is no wonder that Goodwin expressed the merits of both Goethe and Bateson's whole organism approach and their respective science of wholes (Goodwin 2006, 2008). Today, a further significant aspect of postgenomic biology (Goodwin's term) is that genomes and proteomes are now being modeled as self-referential networks, "a description that beckons in a new direction."[1]

ECOLOGICAL TIMING CYCLES

It may come as a surprise to those looking at ecology for the first time that organisms are only one among several criteria that can be used for the study of environment. In the nineteenth century, the study of organisms in the setting of landscape organisms was a dominant criterion, and their study was one of tangible objects in a visual space. One of the best-known ecological expressions in this era was, "Everything is connected to everything else," as object to object, event to event, in a definite manner. This sentiment has persisted even though ecology has moved on to deal with criteria other than organisms and with specific associations or relationships in a variety of ecological settings beside that of landscapes. Thus an ecological study of "population" deals largely with animals spread out in surfaces far wider than those landscape, while an ecological study of "communities" deals largely with vegetation in a variety of surfaces, some very widespread. The choice is left to the observer.

As Timothy Allen and Thomas Hoekstra point out, conventional expression in biology revolves around four paradigms, those of organism, species, evolution, and mechanism, but an ecological study is different, for "ecology" is always tied to the perception of observers and of what constitutes their idea of an ecological association. Ecological studies are expressions of "logical type." Any measurement observers draw from their study as an ecological feature has a great deal of subjective appraisal lying underneath their objective accounting. In addition, ecological studies will vary in scale from local studies of population to studies of biomes that may include studies of climate along with organisms, animal population, community vegetation, and ecosystems. At the scale, or level, of the biosphere, physical conditions of climate are dominant and this "high" level ecological determinant modifies or constrains all lower level activity. Yet only rarely do ecologists find reason to portray the integration of levels as being that of a pure hierarchy, namely one level nested in another level of order, so that a higher level contains and controls another level of order. Ecological phenomena do not constitute a hierarchy as we understand a political hierarchy, especially if one accepts James Lovelock's Gaia hypothesis. Gaia proposes that living systems are not directly controlled in a hierarchical manner by the physical determinants of climate because living systems are able to modify effects of these physical conditions for their own benefit. And the modifications are sufficiently robust to be able to persist in creating a "context" for their ongoing continuance or survival (Allen and Hoekstra 1992, 244–245).

One of the ways in which living systems escape constraints of the physical world, as we have already seen, is by recycling resources. The recycling pathways enable an ecosystem to achieve a sort of identity, or entity-ness through repetitive ordering of cyclic processes. It is this repetitive cycling that constitutes a dominant criterion in the study of ecosystems. An ecosystem emerges as a set of scaled processes integrated in terms of their turnover times in a recycling pathway. Ecosystem pathways develop visible patterns of slow and fast cycles, but their mode of integration is more difficult to observe, and the "context" that recycling creates to modify physical pressures is an even more intangible process. Though the ecosystem pathways of recycling definitely include organisms and organisms themselves contribute to orientation of the pathway, the visibility of separate organisms in the cycling process tends to disappear. Their pattern of connectivity is very different from the Victorian notion of a network of tangible objects. Instead their connectivity is marked through different patterns of relations in recycling processes.

Allen and Hoekstra present a diagram to show why individual biota are not usually considered as distinct slices in these recycling pathways (Figure 7.1). The first aspect to note is that pathways drift in a process of recycling between the biotic (living) and abiotic (nonliving) stages of the recycle. To pass once through the whole cycle, a mineral nutrient must pass in and out of the biotic stage into the abiotic stage several times in one year. Thus the leaf falls from the tree; worms eat the leaf; rainwater washes the nutrients into the soil directly from the leaf and from the feces of the worm; fungi absorb those nutrients and convey them to the root to which they are connected; the root dies leaving a frozen core of nutrients; in the spring new roots grow down the old root hole, collecting the nutrients; the rest of the plant passes them up to the leaves (Allen and Hoekstra 1992, 46).

Connectivity lies in the relation of one set of processes interacting with another set of processes, processes that are not even of the same kind: the physical degradation of the abiotic (or material world) has relational frequency with biotic or activity of the living. It is not simply the physical presence of rainwater that is crucial as it washes the nutrients into the soil directly from the leaf and from the feces of the worm, but the interaction of fungi as it ab-

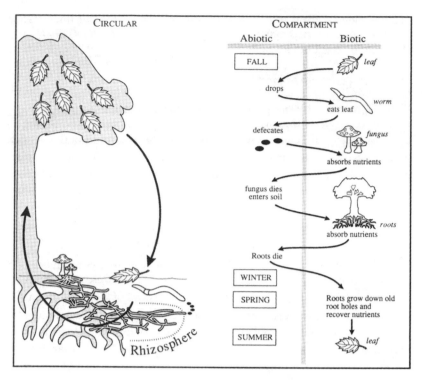

FIGURE 7.1. Mutuality of living and nonliving domains in seasonal cycles of nutrients.

sorbs those nutrients and conveys them to the root to which they are connected keeps the recycling process in motion. The activity of the tree pushing new roots row down the old root hole and collecting the nutrients in the spring enhances potential feedback to maintain the recycle.

An entirely open system would make any conservation of a throughput of resources very unlikely. Recycling pathways introduce a type of mixed biotic and physical closure into what would otherwise be an entirely open system with regard to input and output of resources. In the recycling process biota respond not only to the immediate presence of mineral nutrients, but also to context of their timing and frequency at particular entries into the recycling process through developing sensitivity to frequency of return time. Frequency of return time "stacks" the order of events in an ecological system. "Higher" levels of ecological order are "stacks" associated with longer return times, which is to say, they occur with lower frequency. Lower frequency is typical of processes that are more constant over time, and their lack of change permits "higher" or upper levels to become a context for lower level but high frequency events. If the return time is annual, as in Figure 7.1, return times are very frequent and indicate a "lower" level stack of events.

"Survival of the fittest is in fact survival of the ones that fit the context," say Allen and Hoekstra, an argument that is in the same tradition as Bateson. Sometimes the high level context may involve change, but in these cases, the constancy is that of the type of change that "always happens," for example the El Niño–La Niña phenomenon. Yet it is not always transparent what are the higher, or the lower, or the critical constancies that create a context, because there will always be sublevels in a recycling or recursive pathway that are differently scaled (Allen and Hoekstra 1995, 31).

If the closure and conservation of resource is dependent on the activity of biota enhancing feedback potentials in the recycling processes, and if the scaling of ecosystem pathways embed sensitivity to context, then such responsive interaction between organism and environment suggests that there are information pathways along with physical pathways in ecosystem. Historically ecologists who have investigated information feedback potential of biota have presented evidence of information as feedback in terms of mechanistic interactions; for example, biota derive their information from the biochemical mechanics of signals, input transducers, decoders, channels, and nets, they claim. Only a smaller number interpret the presence of information in ecosystems in terms of organisms having embedded awareness as a result of their own perceptions providing an orientation either for change or for constancy.

For these few authors, all organisms rummaging in their environment around display "meaningful activity." A key person advocating the latter view of interaction between organism and environment is Jacob von Uexküll. "Meaning is the guiding star that biology must follow," he wrote, and every action of an organism can be shown to relate perception cues to some effector. Even when the grazing cow transforms a flower stem into wholesome fodder, it imprints its meaning on meaningless objects and thereby, in this simplest act, makes a flower stem a subject-related "meaning-carrier." Biology should acknowledge that there exists a functional circle between preceptor organs and effector organs in which objects in the form of meaning-carriers connect the meaning-carrier with the subject. In short, physical entities do not exist simply as physical entities in interaction with other physical entities, but also exist as perceptual cues to which organisms learn to respond in the course of their existence (von Uexküll quoted in Favareau 2010, 94–95).

Von Uexküll talks of an animal's perception as "Umwelt," or "self-world," and wrote that here lies a functional connection mediated through guiding signs between the animals "innerworld," and the "outerworld" of environment. Today we would call his term "outerworld," an "ecological niche." Guiding signs serve the animal's orientation to a number of activities (the presence of food being one) that the external world presents to the animal's perception as a "meaning-carrier" of a selected activity. During a lifetime of an animal, many other perception-motor connections are learned through experience. An animal's behavior when confronted with a certain specific stimulus would act in response to that stimulus in terms of what a situation means to the animal at that moment in terms of its survival. He was making no attempt to read any animal's mind, nor was he talking about animal reflexes in the style of behavioral psychology. His function-circle was a very credible example of an organism's activity in relation to feedback cycling in a circuit of events, long before the cybernetic ideas of systemic feedback had become established.

His classic study was of the wood or deer tick, a study already referred to above in the previous chapter. The wood or deer tick was merely one of many studies von Uexküll made the relation of organisms to "meaning-carriers," but his study is special because the tick is particularly devoid of the usual array of bodily sensors, for example it has no ability to taste. Reviewing the number of studies that he undertook of the Umwelt of various creatures, von Uexküll wrote that meaning-carriers—in the tick's case butyric acid—are "secret signs" or symbols that members of the same species can understand but that other species cannot comprehend. In the case of the tick, there are only three meaning-carriers that the tick can respond to as it hangs in a tree, waiting for an animal to pass under it. Yet, to succeed, it must respond to each of these meaning-carriers in correct sequence. The first of these is butyric acid emanating from the skin glands of a mammal below; the second is the shock of hitting the hairs of the mammal, which stimulates a response of running around; and the third is sensing heat, which stimulates its boring response into a warm membrane. If the tick lands on a something cold, it must climb back up the tree. The tick, lacking a sense of taste, will puncture any warm membrane. A second, more astonishing feature then appears in that "each subject's [organism's] symbol is at the same time a meaningful theme for the structure of the subject's body" in other words there is an association between the structure of the organism's perceptual symbols and the body plan of its effector organs. There is a point and counterpoint relation between the organs of perception, the relevance of cues and signs available to the organism in its particular habitat, and the actual composition of the organism's bodily structure. The tick's body plan expresses certain themes distinctive to all mammals, butyric acid being a counterpoint to the shaping of the body form of the tick.

This point, counterpoint relation, is even more remarkably demonstrated in organisms whose activities complement one another, expressed for example in the honeybees "flower likeness" and in the "bee likeness" of the flower (von Uexküll quoted in Favareau 2010, 106–107). By means of his initial distinction of Umwelt as "self-world," von Uexküll's musical metaphor of point and counterpoint proceeds to configure holism in nature. We will meet von Uexküll again in the following chapter as one of the fathers of biosemiotics.

Allen and Hoekstra do not produce a methodology about "meaning-carriers" in a field of ecological events, but they do acknowledge the subjectivity of observer's choice and they provide a sophisticated approach to the process of their observing movements through ecosystem levels, aligning observers' subjectivity with the patterns of ecosystem events. They devise what they call a "layered cake" model as a template for the study of ecological systems. The layer cake they describe is one in which the observer is able to move across multiple criteria. Upper levels of any scale in the stacking process, according to any criterion, define role, purpose, and boundary conditions of an informational order; lower levels may produce mechanisms useful as explanatory principles of a material order. In other words, they invent ways in which an observer's mapping can undertake rapid and easy shifts between nested levels of an ecological order, which in turn respond to information processes.

They also propose a topological approach to ecosystem representation. They suggest that ecologists represent a field of cycles like a hollow tree log in which the outer "bark" of the log is composed as a set of cycles passing through recursively the "top" and the "bottom" of the tree. Many of these cycles are intangible, such as the birth-death cycle of the tree. Other cycles are depicted on the outer "bark" are timing cycles that the tree shares with other trees and other biota. When this "tree" or reference structure is depicted, it appears as a topological torus. Seen in this respect, a living tree is a series of temporal processes flowing through merging atmosphere, ground water, birds, insect life, fungus, and bacteria from other living cycles into its local organization. The trunk of the tree is rendered as the "hole in the bagel" through which multivalent processes pass, in other words a hole for the patterns that connect. In fact, its recursive field of connections is so similar to that of a bagel or a doughnut, that the two authors quip that ecology "is just a bunch of bagels." Its recursive ordering gives a sense of unity and of mutual coordination that would be difficult to try to fragment or reduce for purposes of analysis.

More recently, Allen has come to the conclusion that differences between the "realist" definition of scientific enterprise and his multiple-perspective approach invites a push toward a more subjective stance in the whole of ecological science, by placing supposed "objective" practices of measurement within the wider frame of narratives drawn from scientists, stakeholders, and the

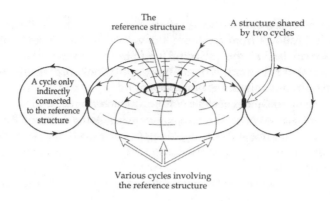

FIGURE 7.2. Allen and Hoekstra: The recursive topology of ecological structure.

public, in any research undertaking (Zellmer, Allen, and Kesseboehmer 2008, 171–182).

Allen agrees with Bateson's viewpoint that the introduction of subjectivity does not create junk science. On the contrary, the subjectivity of narrative adds to science; it produces a different logical type that can play a positive role in the evaluation of any scientific project. Instances of subjective–objective complementary occur when the scientific project is called upon to conduct measurement, and other aspects of calibration to improve the quality of a narrative. The scientist, as a participant in the process, can at any time raise the quality of narratives used by challenging each percipient with calibrated results from his or her models of experience.

ECOLOGICAL FLOWS AND CONNECTIVITY

Topology is the mathematicians' name for studies of processes in systems that entail connectivity and closeness. The topology of a donut or bagel can be modeled by taking a flat piece of paper, rolling it over and over, then joining the rolled over end of the paper together in a single unit. Topological closeness is not a matter of distance as in geometry, but depends upon the immediacy and directness of connections—inner and outer, within and among, perceptual and actual. These topological curlings remain unchanged in their domain of connectivity despite any stretching or enfolding of surfaces, regardless of continuous transformations. Topology deals with constancy, connective inclusions, continuity of connection in turbulence, overriding the incredible variety that otherwise can be found in forms, shapes, sizes, locations, dimensions, and patterns.

Allen and Hoekstra's notion of ecology as "just a bunch of bagels" is intuitively correct, according to Don McNeil,[2] but their topological representation of interconnection of organism and environment requires a more detailed elaboration of its torus-like features. For ecosystems to be depicted topologically, the "bagel" not only must have recursive circularity but also must enable differential timing and other sorts of change and adjustments that occur in ecosystems to be depicted in a more complex manner. The bagel's "patterns which connect," the recursive curls that close back on themselves, must allow cybernetic patterns depicting mutual entailment in respect of each other. They must also enable depiction of topological binaries, in that ecosystem movement is not a continual stream of integration because some pathways in their oscillations will inevitably counteract each other, and the appearance of binaries forming from processes of mutual entailment is inevitable. Finally, the topology must depict a more accurate representation of "part" and "whole," for the hole in the bagel is a crucial composition in its unity. This asymmetry of the bagel enables the depiction of a far more complex structure than that available in the diagrams of spheres, planes, matrices, pies, and projective cones that pervade scientific representation of process in ecosystems. None of the latter enables an appropriate appreciation of the intangible connections of part and whole in holism.

The dynamic stability of living systems includes many, many swirls and streams and curls modifying their echelons of order. But McNeil notes that only relatively very few organized and persistent "eddies" are able to maintain themselves in the general turbulence of living systems. Where such "eddies" do occur, however, they can be extremely robust and achieve a *relative invariance*. A flow that keeps itself "going on" and persists in living systems has a general characteristic that its gyrations counteract excursions in its circulations so that they do not dissipate and disappear. Any flow that maintains itself and is balanced in its oscillation has a circulation that is nonlinear and bidirectional over time because its recursing flows are cybernetic. That is to say, the feedback in the system permits increase until decrease is required and decrease until increase is required and so on indefinitely. This feature may be called the "virtuous cycle" of a relative invariant, and is contrast to a "vicious cycle," where the pattern of recursing flows becomes destroyed through more leading to more or less leads to less until their final deformation. In an ecosystem, one of the better strategies is to find the intersections of flow and to trace the "virtuous circles" in relation to the "vicious circles." Borrowing a phrase from Heraclitus, McNeil notes that the relative invariance of virtual circles is one in which all the cycles in that circle are able to "rest through changing." In other words, our own bodies are stable relative invariants, but they also flush out and change their whole composition once every seven years.

As we have seen, timing or frequency is crucial to maintaining connection in the flow of events, and a characteristic of timing is its oscillation. At any intersection of events, a counteraction may yield less, or in another instance a counteraction to less may result in a condition of less leading to more. This "enoughness" that emerges in persistence is a qualitative form that accompanies cybernetic flow, and occurs because the reference regulators of one flow have interaction with the reference regulators of another flow in ecosystems. The goings-on of "enoughness" result in convergence that, in another cycle will lead to divergence, and vice versa. So long as the cycles of change are organized to compensate in their oscillation for their own deviations and departures, the constraints of "time" become an aspect of "timing" rhythms, an outcome of persistent circulation that continues to produce relative invariance.

Cybernations maintain themselves through maintaining relatively stronger connections than if they were controlled by a superordination, for the superordination occurs less frequently and has weaker bonding with its lower echelons. Nevertheless there are forms of stratification and appearance of discrete levels in such structures. Ecological structures are not a series of smooth gradients. The pathways of relative invariants tend to create niches and stratify themselves into echelons. Such stratification, though discrete, does not stop mutually supporting flows, rather the lower echelons may supply material for the cybernation of higher echelons; while the higher echelons may set constraint references or context for lower echelons, thus maintaining the organization of the whole system. Complementarity of this sort yields a concept of heterarchy. Warren McCulloch's idea is repeated in the higher and lower frequencies of recycling, which generate timing contexts within the pattern of pathways in an ecosystem.

Heinz von Foerster picked up McCulloch's ideas in his role as secretary of the Macy Conferences on Cybernetics. He later produced a model of how flows might create mutual support in cybernation as a result of one flow becoming a reference set of constraints for the other, and vice versa, both creating and maintaining each other. He produced a diagram of two areas of circular symmetry merging through overlap of doubly closed circuits lying on top of one another, as in a donut or bagel. The model was of neural circuits and of endocrine circuits lying in circular manner on top of one another, as might occur in a "sensory motor synapse" of a cell. In the complex circumstance of "fight or flee," the two circuits could be binary so far as effector activities are concerned yet remain complementary in their overall composition. "This is the organization of a biological entity in a nutshell—or, if you wish, in a doughnut shell," von Foerster remarks (Segal 1986, 134–135).

The top circuit of the bagel is known as the *annular* curl, while the "thickness" of the roll, which travels from top to bottom of the bagel's surface, is called the *meridial* curl. As von Foerster showed, the meridial cyclical orderings alter the parameters of the annular cycles, and vice versa in such a way as to stabilize the whole with mutual and complementary control. This occurs even though the two curls in other contexts could in some circumstances be depicted as binaries. The important topological feature is that in virtuous circles the interactions of the neural and the endocrine take place without losing their robustness and going out of control.

McNeil first met Heinz von Foerster at a meeting of the International Society for the Systems Sciences (ISSS) in 1994, where he had given a paper on toroidal heterarchy. Von Foerster gave a "hell-raising plenary sermon" on how Warren McCulloch's paper, "A Heterarchy of Values in Neural Nets," remained unappreciated. As McNeil told me,

> He waved a copy of it at the audience in what today would be considered a threatening and politically incorrect manner. Toroidal heterarchy is, of course, a theme of McCulloch's paper. Everyone in the room who had made fun of my paper and the topological imagery therein turned toward me as if in awe and apology, and I modestly accepted the implied vindication. But everyone was back to their old prejudices in favor of "hierarchy" by the end of the subsequent coffee-break. My fifteen minutes of fame that day certainly did not last.

McNeil shows how the apparent empty space in the hole of the bagel becomes that third dimension of a bagel necessary for heterarchical interaction. He introduces a third and fourth curl as qualitative conditions that can be expressed in the apparent empty space of a torus. A "hole" usually indicates an absence of morphological form, but, in a bagel it is an important configuration, one recognized many centuries ago as the "Axle and the Wheel" configuration. Axle and wheel combine so that the hole in the axle holds together spokes as they travel along the ground while the hub enable the spokes to turn on difficult ground conditions.

The *Daodejing* (403–221 BCE) is one notable treatise whose discourse about the axle and the wheel suggests that mutual entailment can occur in spite of its apparent "nothingness" and apparent opposition of "presence" and "absence" of filled space. Thus:

> The thirty spokes converge at one hub
> But the utility of the cart is a function of the nothingness (*wu*) inside
>   the hub.[3]

An axle and wheel is a very material example of mutual holding, but McNeil's depictions are nontangible examples of the phenomenon. He argues that while the hole in the donut appears to be absent space, it can contain two spirals or "twists," called the Villacreaux curls. These curls can either spiral right spiral or left spiral. They are a "presence" that occurs for example in the updrafts and down drafts of a whirlpool, or in the dynamic of the eye of a hurricane as it either dissipates or enlarges over bodies of water or land. The Villacreaux twists produce a flow that can travel "upward" as an *inside* flow of the donut but then can overflow at the top and move "downward." A Villacreaux trace spirals from the center successively upward and outward, then outward and downward, downward and inward, then inward and upward, hence back to the center, thus to embody a living, breathing metaphor for holism in biota.

As Figure 7.3 shows, the three dimensions are fully engaged in a topology of "re-spiration" which occurs as a result of continuous oscillation in one direction, going around, and through, and then turning upon itself. In the figure, the annular curl is depicted in red; the meridial curve is depicted in blue; and the Villacreaux tracing is depicted in yellow. Note that annual and meridial curls are depicted as potential binaries in the form of "analog" and "digital," but all together they are a heterarchy of mutual complementarity in what McNeil calls "goings on."

The "nothingness" of hole in the bagel permits change and adjustment of the polarity of the various cycles and hence enhances their mutual complementarity. Here, the order of events is important. Order matters at the core of all that is going on, and the recursive cycles must proceed in order from their creation to their excursions and to their counteraction. The sequence

Perception Embedded in a Heterarchy of Goings On

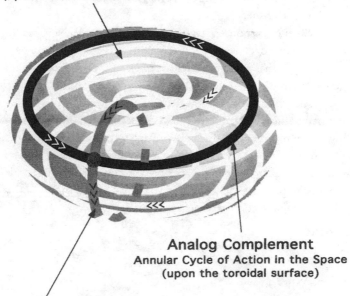

Vantage
Helical Villacreaux Cycle
of Perception and Interpretation
from *within* the Space of Action
(upon the toroidal surface)

Analog Complement
Annular Cycle of Action in the Space
(upon the toroidal surface)

Digital Complement
Meridial Cycle of Action in the Space
(upon the toroidal surface)

FIGURE 7.3. Perception embedded in a heterarchy of goings-on.

of "nextness" is of concern. The continuance of mutual complementarity is dependent on continual oscillations where more leads to less, and less to more; or sequences where convergence leads to divergence and vice versa. Going out of turn can be a prescription for running amok, for a counteroscillation before an excursion is itself an excursion that can lead to instability in the sequence of oscillations. Thus, cybernations of virtuous circles develop flows through sequences in feedback that themselves develop appropriate sequences as a means for keeping order.

The potential connectivity in the topological form of a torus is, in fact, amazing. Mathematicians have estimated that a torus cast into a knot and that initially has sixteen crossings of a closed curve in space (McNeil works with toruses that have seven to twelve such crossings) can take one of 1,388,705 permutations that do not intersect themselves anywhere. The first big question for percipients to reconcile in an ecosystem is to decide what is included in "the *system*" and what is excluded from it, in "the *environment*." Ultimately, percipients must decide this for the purposes at hand, but if they do not do so in accord with what is going on, regardless of their perceptions, all manner of mischief can ensue.

Percipience is at once imaginary and real, says McNeil, real insofar as relative invariants of cybernation that keep things going on can be observed; it is imaginary in that a percipient will define its inclusions and boundaries, and the percipient inside the system is likely to define systemness and coevolution and heterarchy differently than the percipient observing from outside the system. Human beings also run into the difficulty of visualizing spatial dimensions higher than three, and are therefore apt to reduce higher dimensions to the formalism of three. Yet, picturing a torus beyond three dimensions is not only possible but also necessary in ecosystemic contexts in order to depict the phenomena of part and the whole and the-one-and-the-many. Both are crucial to a holistic appreciation of an ecosystem.

8

TOWARD THE SEMIOSPHERE

We now switch from observers' perception of living systems to creatural perceptions. This chapter presents two areas of research that have provided advances on Bateson's enquiry, one of which goes under the name of biosemiotics, which acknowledges Bateson's writing as a precursor of its own approach. Biosemiotics is currently focused on the "Copenhagen school" led by Jesper Hoffmeyer, a molecular biologist and writer on science for the public at large (Hoffmeyer 2008b). The other, biosemantics, has developed its discourse with no particular reference to Bateson. Biosemantics largely follows the writing of the biologist and philosopher Ruth Garrett Millikan, whose discussion of the "varieties of meaning" provides both empirical and philosophical support for the semantic mapping of animal experience.

Biosemiotics proposes that semiosis (patterns of meaning and sensibility in their broadest sense) underpins coherence in all living systems, and further that living systems organize and stabilize around a capacity to perceive, to communicate, and to learn. Biosemiotics has a vision of life that is one of organisms and organic processes responding to themselves in relation to others, especially about change in the natural world. It has incorporated both Jacob von Uexküll's notion of Umwelt and Bateson's vision of the world as Creatura, where species of animals, insects, and plants have a complex capacity for perception and interpretation. It claims there would be no living system at all if physical signaling to-and-fro was all there was to receiving a communication. It holds that information in life processes is preeminently reflexive and that self-reflexive information evokes a propositional ordering, compared to mere signaling between biochemical "bits." Response to a sign involves more than an automatic reaction to a signal. That is to say, for a response to occur, the signal must also include propositional information "about" information to inform, or, in Bateson's terminology, provide sufficient redundancies or patterns for the context of a sign to enable meaningful response.

Organisms not only belong to ecological niches in the physical sense, but also to *semiotic niches* whose cues are visual, acoustic, chemical, and the like, and that must be correctly interpreted for the individual organism to survive. The vocabulary of biosemiotics looks to the semiotics of C. S. Peirce for appropriate terms that capture the highly flexible semiogenic capacity of living systems. As a subdiscipline in science, biosemiotics gives special recognition to the work of the disciple of von Uexküll, Thomas Sebeok, and his circle of friends, who promoted zoosemiotics and animal ethology in the mid–twentieth century and who moved those disciplines toward a more holistic study of species' responses to environment. This integrative work ended with Sebeok's death in 2001 and with a handing over of his conceptual theories to a group of scientists in Denmark and other Nordic countries, the core of today's practitioners in biosemiotics.

Biosemiotics argues that because of living systems widespread semiotic capacity to interpret in the biotic world, natural selection is never blind and is always guided by semiotic integration of prior species. In its widest reach, biosemiotics contends that there is a "semiosphere" as enveloping as "the biosphere," that follows a "semio-logic" of communication, which biology has ignored but is as real as the biosphere. The broad notion of the semiosphere is credited to Juri Lotman, a Russian-born Estonian. Hoffmeyer argues that the semiosphere incorporates all forms of communication, sounds, smells, movements, colors, shapes, electrical fields, thermal radiation, chemical signals, and waves of all kinds. Biosemiotics proposes that "the biosphere must be viewed in the light of the semiosphere rather than the other way around" (Hoffmeyer 1996a). In a semiosphere, living creatures react selectively to their own environment and coevolve through shared meaning with other living forms in the same ecological niche, an interrelation that evolves over time.

The notion of a semiosphere defined in this manner considerably advances Bateson's ideas about the presence of communication in all living systems, and of "mind" as an aspect of living systems. Hoffmeyer's vision for biosemiotics is to develop a methodology that will prove to be more acceptable than biology's current conceptions of information. In this sense, biosemiotics follows Bateson in challenging Shannon's postulates of information with no capacity for "interpretation." Biosemiotics is postgenomic and aligned with DST as well (Hoffmeyer 2008, 372–374).

BIOSEMANTICS AND MEANING RATIONALISM

Generally speaking, humans use language as a vehicle for transmission of theoretical judgment, and use thought for attaining their purposes or goal states, says Millikan. Thought and theoretical judgments then mesh, through rational calculation, with their explicit purposes. It would seem to be unreasonable to apply the same stream of decision making to the "mind" of non-human animals, yet a large number of Western philosophers hold views in concord with this set of conditions. Both Western philosophers and Western scientists have taken rational purposes to be a primary criterion for any presence of mind, even if some are not fully aware of their presuppositions about this question. Ruth Garrett Millikan calls this mistake "meaning rationalism." Meaning rationalism, Millikan remarks, still "permeates nearly every nook and cranny of our [Western] philosophical tradition" (Millikan 1984, 91, 92), and its mistaken approach to criteria of "mind" or of "meaning" has lasted far too long. Meaning rationalists hold that meaning occurs only when conditions of intentionality—in the philosophical sense—can be shown to exist. They hold that meaning arises from the conjunction of three concepts—intentionality, rationality, and consciousness. Without the presence of all three criteria, there is no evidence of "mind," and, therefore, no evidence of meaning. Thus according to their strict definition nonhuman animals (let us use Bateson's term "creatura") are nonminded.

Common examples of meaning rationalism lie in using indices borrowed from game theory to explain predator-prey relations. Other examples include the use of utility functions to define effectiveness in creatural mating behavior; that is to say, explaining mating behavior through various forms of cost-benefit analysis. Yet another example is the way in which creatural adaptations to environment become analyzed in relation to energy efficiencies, always a seeming requirement for rational behavior. Meaning rationalists assume that since a particular set of conditions demonstrating rationality holds true in human activity, these conditions must be expected in creatural consciousness as well.

In Millikan's view, the inherent physicalism of their approach traps meaning rationalists in a truncated view of the interrelation between humanity and nature. Meanwhile, the actual processes through which creatural interpretation occurs and affects behavior go unnoticed, along with their enormous complexity. Millikan states categorically: "I claim that rationality is not the mother of intentionality" (among nonhuman creatura) (Millikan 2004, 84). Therefore, a discussion of creatural meaning and intentionality must begin with different initial premises.

Her discourse begins with the notion of a "natural sign" and proceeds to explain how use of natural signs is a valid way of expressing meaning and intentions. Like Bateson, Millikan's epistemological project aims at showing how it is possible for an animal to recognize the recurrence (recursion) of a natural sign and use that sign effectively for its own purposes (Millikan 2004). Philosophers and biological science are loath to use the concept of "natural signs," but as evidence of creatura's use of natural signs becomes available, usage of natural signs by creatura can be compared in an overall way to the use of language among human beings.

Sign usage is linked to perception, and some aspects of sign perception among and between creatura bear a strong resemblance to the way in which human beings engage in their own perceptive activity. At the same time, how does an animal know that it is doing all right in a run of a day, when natural signs—even of the same thing—are manifested through diverse media under a wide variety of conditions and through differing sensory modalities? Millikan first defines a natural sign of a thing or event as something from which creatura can learn by tracking a connection that exists in nature. Among human beings, we would probably say that the connection can be tracked through linguistically generated "thought" (see the discussion of abduction that follows). Yet, the use of natural signs by creatura is entirely distinct in an evolutionary sense from the way in which language use became developed among human beings. Moreover, we know little of "thought" in the creatural world, especially in the case of simpler forms of animals. What we do know is that perceptual systems of creatura are deeply dependent on local information in nature. And the most important criterion for the use of natural signs by creatura is that the natural sign be a "locally recurrent sign."

For an animal to interpret a locally recurrent sign successfully, it must keep within its natural domain and stay within its boundaries, and by this means creatura can use natural signs to collect information about the world. In order to "mean," a naturally recurring local sign must be strong enough that relations between tokens of the sign and the affairs signified are reiterated in that domain, and that this reiterated relation between the sign and the affairs signified each persist in that same relation to one another. Yet, it is not necessary that there be any determinant causal connection between a locally recurrent sign and the affair it signifies. "Recurring" in the phrase "recurring natural sign" does not necessarily mean that the exact same sign recurs over and over again, but rather that the same sign-signified relation recurs. The minimal applicability of the sign to a certain natural structure would be that a natural sign includes aspects of its own time and place. Put another way, a "recurrence," in the case of recurrent signs, refers to signs in a system of signs, where a sign system consists of a set of sign types, each of which pertain to recurring relations between the communicators. Millikan calls these "semantic mapping functions" (Millikan 2004, 49–50), and they are strongly reminiscent of Bateson's notion of "context" derived from feedback. Her sign-signified relation is recalled with respect to distinctive patterns, and like Bateson's notion of context, Millikan's semantic mapping functions are layered as to time and place.

Millikan then adds the notion of "intentionality." An intentional sign is not just an iconic representation of some aspect of creatura's environmental domain; it is also a sign for some kind of interpreter and interpretation. Intentional representations are designed to have the effect of correctly representing "something" arising out of some form of perception or communication to some other organism or interpreting system. Intentional signs are signs "apt for use" by sign-users. No one can doubt that when a hen clucks it has a definite effect on her chick. The cluck call is not merely a "picture" of a representation of food that happens also to be used by the chicks. With the cluck call, the hen has purposefully produced a sign of food for the chicks that, like any natural sign, a locally recurrent sign, is "intentional" in the common sense meaning of that term—it calls the chicks. An intentional sign must obviously induce cooperation within interpreting systems, such as a clucking hen and listening chick.

Millikan argues that such a sign meets the criterion for "intentional" in the philosophical sense as well, for philosophers define intentional representations as coordinating sensibility, perception and communication in purposeful activity. For meaning and intentionality to arise, it is only necessary that the sign both be a sign that recurs, and a sign that is used by the communicators—both utterers and interpreters—when each is closely related to the existence of natural signs. In addition, the signs used in communication must constrain the relational behavior between utterers and interpreters so that "utterers" of the sign-using system in some way coevolve with sign interpreters. Communication feedback then generates a correlation between similar signs and similar signifieds that occur within a natural domain of time and space (Millikan 2004, 40–47). As long as a locally recurrent sign is kept within its natural domain, appropriate interpretation continues to occur.

It is not necessary that there be any rationally causal connection between a locally recurrent sign and the affair that it signifies. Millikan, like Bateson, agrees that communicative situations in local creatura domains lie somewhere between the propositional structures of language used by humans and random trial-and-error behavior. Given that they are events or occurrences in a local state of affairs, the relation between tokens of the sign and the affairs must keep specifying that relation. Thus, Millikan's "natural sign" can be learned and acted upon through perception, which gives rise to distinctive perception-action loops or feedback, which in turn yields purposive behavior. Hence, for the observer, a broad notion of recursion becomes apparent.

The importance of perceptual appraisal, compared to conceptual appraisal, is a pervasive characteristic of Millikan's analysis, as it is with Bateson's. A squirrel does not conceptualize its purpose when trying to jump from a branch in a tree to a place where nuts are stored, nor does it ever make strict logical inferences about this situation. Rather, the squirrel studies the perceptual situation at length, first from one angle then another, and experiments through trial-and-error until it "sees" a way it might try to jump. Its own perceptions enable a sort of perceptual trail about behaviors in occurrences or states of affairs that may or may not yield a purposive conclusion (Millikan 2004, 12). Perception among creatura is a way of understanding natural signs, or, better, of translating natural signs into intentional signs. In other words, sign-use is built on boundaries, and levels and contexts and repetitions (or to use Bateson's expression, "redundancies") that occur with multiple levels of perception, both distal and near.

Both humans and nonhuman animals relate their percepts to immediate action. The "propositional requirement" for use of percepts is simply that the same kinds of sign are connected through various relations to the same kinds of signifieds, and that they recur with the same natural conditions of signification or meaning. Yet, when humans learn through language, meaning is not typically grasped in this way (Millikan 2004, 122). This indicates that a sequence of events in creatural perceptual "tracking" is neither congruent with the selection procedures that lie behind human intentionality nor congruent with the selection procedures resulting from skilled learning (Millikan 2004, 12–13). Instead, perceptual tracking yields perceptual-action cycles, essentially through a trial-and-error process. By this process, the perceptual awareness of their own sensory signals among creatura generates a level of self-reflexiveness and "purposiveness" quite separate from those designated in human rational-purposive activity.

If success at appropriate tracking is an important means through which creatura learn, learning through perception requires memory of the sequences of events that led to success. Some creatura are quite limited in their recall of these sequences, while others are able to construct a representation or map of a territory that they have explored and, for example, are able to remember a very large number of places in which they have cached food. Some even seem able to map a general-purpose spatial layout of their territory and are able to use this spatial representation in relation to more distant places in which they undertake activity. In most cases, these semantic mapping functions are not linear, so it is often possible for creatura to recognize distal signifieds without recognizing all or any of the signs on the route. There is always a means for semantic mapping that goes directly from a sign of a sign to the more distal affair that it may signify. This, of course, is a human capacity as well—for example, human beings register only digital patterns on our retina, but perceive colors and shapes.

Put another way, context allows creatura to perceive possible analogues or isomorphisms that exist between the set of possible signs in one certain domain and the set of possible signifieds in another domain, and this gives real flexibility to their communicative behavior. Yet, it is usual for creatura to perceive only what they have motivation to exploit at that particular moment.[1] In some other instances, creatura can be guided toward a future by a perception of that which will occur later. But Millikan suggests any end state is unlikely to be represented, or represented articulately in their "mind." There is no need to think of creatura as perceiving everything and everywhere about them as possibilities for action (Millikan 2004, 87). Their activity is of a different kind from procedures of conceptual thought that human beings favor, yet the evident differences seem to have been passed over in conventional biology.

As with Bateson (2000, 51–56), Millikan acknowledges that lack of a negative form of expression (a "not," or a "do not") is a crucial limitation in creatural sign use. Human language is both descriptive and injunctive (Bateson's terms were "report" and "command") in a positive and a negative form, and there are clear differences in the way in which human beings use description and injunction as forms of communication. Animal communication by comparison is solely positive and solely injunctive. The rabbit thumps its hind feet smartly to the ground in order to trigger the injunctive response of "Take cover," which results in other rabbits suddenly stopping and then taking cover (Millikan 2004, 89). But such communication possibilities do not permit "don't take cover until the count of three." Among humans, an overwhelming use of language pertains to description, or descriptive terms modifying injunctive commands.

A more complex example is the wagging dance of the honeybees. Here is a sophisticated use of signs by worker bees to indicate the direction in which other bees in their hive can find nectar. Yet, the honeybees have no way of saying that the bee must follow the sun for a particular time as it lies at a certain angle between the hive and the nectar. And there is no negation transformation of the dance that says, "Don't go over there, there is no nectar." The mapping and sign usage of honeybees is inarticulate even compared to muddled sign direction that constantly occurs in human language terms. Honeybees have to learn independently to which local domain the waggle dance sign of the honeybee is pointing, for the wagging dance itself does not say (Millikan

2004, 91–93). There are other evident properties typical of human language use not apparent in creatura's semantic mapping, and some of these arose in Bateson's own research on cetaceans.

Questions of meaning rationalism arose in Bateson's research on dolphins, especially in the case of dolphins trained to perform for human audiences. It was a task that he had taken up with John Lilly upon leaving Palo Alto and his former colleagues. Much of his research on cetaceans was motivated by the idea that dolphins would show behavioral characteristics of a higher order than those of other mammals. Yet, a prime difficulty in all of this work, he wrote in his *Notebook*, was that of conceiving what was meant by the word "higher." Most of the available data in Lilly's research station was behavioristic and had been produced through watching input-output behaviors of the animals. All of Lilly's work evoked ideas of conscious decision making, and by implication, meaning rationalism, none of it seemed to be deduced from the basic principles of evolution and communication theory, with only the slightest reference given to the question of "levels" of their functioning. So while it might be possible to show that these animals can solve complex problems, Lilly's approach made it impossible to evaluate the steps by which such solutions were achieved.

Bateson was working on the notion of "context markers" at the time of his research. A context marker, he told Lilly, "is any percept by which an organism receives information regarding the contingency pattern of the situation in which he is placed. Such a marker, at one end of the scale may simply be ostensive . . . or the context marker may be analogic. . . . At the other end of the scale, the context marker may be purely digital" such as the word "walk" uttered by the master. He quickly pointed out, "the name of the "walk" is not the thing named" (MC, 25/03/1966, 858–96a). The "name" is the relationship "you and me at this moment."

The question for the observer was to watch redundancies in behavior, locate possible context markers for that behavioral redundancy, and record each context redundancy. So where the system indicated to the observer some aspect of context marking in the animal's behavior, the observer should note that segment of experience and also note how the animal maps such segments on to the rest. The expected redundancies would cover behavior in dolphin activity such as mother-child relations, food getting, group swimming, and play. "Thus look at group swimming / look at mother-child relations / look at play . . . and then see how much of the patterns of behaviour are shared" (*Notebook* 38, 1967).

The methodology Bateson proposed here was drawn, at least in part, from earlier work that he had carried out on octopuses. He had argued that the philosophers and ethologists' concern with positivist assessments about how animal "intention movements" related to "consciousness" was phony. The point was that animals could not give clear, precise, positive and negative responses to their "intentions." Instead, octopus discourse centered almost totally "on matters of a higher order abstraction by probing the question "what shall be the styles and rules of our relationship?" To demonstrate "I shall not hurt you" an octopus will expose vulnerable parts of his body to attack, or, alternatively move through several sequences of communication indicating hostile movement. Two octopuses, starting from mutual hostility, will then pass through a sequence of minor battles in which neither gets hurt much at all. After this, the slightly stronger octopus "very slowly and gently embraces the weaker. . . . Following this, the weaker comes over and attacks the stronger with his vulnerable backside—a response to which the stronger retreats i.e. the weaker has now said 'Yes, I know you are not going to attack me' and the stronger has said 'That's right.' The same sequences of hostility exchange indicating peaceful relations (the contrary) could be observed in human ritual as well" (MC, 25/10/1962, 1039–10b).

He told Lilly that his new article on context markers would turn information theory of the original first-order cybernetics "upside down" and make what the [electronic and computer] engineers call "redundancy" in signals, but which he called "pattern," into the primary phenomenon rather than information content. But, he warned Lilly, "it needs to be married off to corresponding hypotheses about what happens in the brain."

CODE-DUALITY

One of Hoffmeyer's most original contributions is expanding Bateson's notion of "code-duality," that is, the way in which organisms enable incorporation of the present into the future through evolution. The concept of code-duality stems from Bateson's observation that life—natural or cultural—depends on the interactions between multiple kinds of coding, and not on genomic coding alone. Bateson basically distinguishes two kinds of coding, originating from different types of feedback. One type of feedback is based on a graded response (such as a governor in a steam engine), and another type is based on on-off thresholds. This is the dichotomy between *analogic* systems (those that vary continuously and in step with magnitudes) and *digital* systems (those that have the on-off characteristic) (Bateson 2002, 123; Emmeche and Hoffmeyer 1991).

Bateson originally created his distinction between digital and analog as being a distinction between *digital*, as codes of numbers, letters, and the like, and *analog*, as attachment to the context of relationship in which it (the message) is given. The distinction is necessary in order to show that both human and animal communication systems are deeply dependent on multiple settings—both paralinguistic in terms of human communication, or nonlinguistic, but strongly relational, in all creatural biocommunication. Human forms of paralinguistic communication occur through "body language," and "body language" always gives humans an evident "double description" in any form of communication with their communicants. Animals have to convey body-language relationship and insight (self/other reference) in a different form. Analog communication is linked to self/other reference through the ability of animals to formulate their own analogies, which they draw from experience. Broadly speaking, animals, even small organisms, learn from the experience of recognizing pattern and are then able to identify in their perception and substitute variations of forms of that pattern in their perception. A concept of double description is required in order to capture aspects of an organism's response to perception. (Plants require separate consideration, and though they can be said to "communicate" [Baluska and Mancuso, 2009], they lack the neurology of the animal kingdom.)

Hoffmeyer also takes Bateson's observations as markers for a discussion of *endosemiotics* at the level of cellular response to pattern. He observes that the most common meaning of a gene in molecular biology is as a single level "reading sequence"—that is a sequence of nucleotides that is transcribed into one piece of messenger RNA that is either translated into protein, or used directly in the metabolism of the cell. As discussed in the previous chapter, the central dogma of molecular biology proposed that genetic information is passively passed on from DNA to proteins according to more or less unambiguous rules. While such an understanding underlines the *replicative* property of a gene, it leaves the *phenotypic* effects of the gene, including all the epigenetic effects that result in the development of a viable organism, totally unaccounted for. Hoffmeyer points out that more recent findings have shown that the transcription of genes to mRNA is in fact highly dependent on the presence of a number of protein factors that themselves reflect the cellular or organismic context in which the transcription process takes place (Hoffmeyer 2008, 78).

He joins with Richard Lewontin and others in developmental biology who put forward the view that genes are quite passive, and on their own almost inert, until brought into relation with the developmental system as a whole. The active components of the cell are the proteins that work together in the subcellular complexes or membranes that are the real "doers" in cellular life and constitute the cell's agency. Code-duality addresses the issue of how apparently arbitrary sequencing results in a coherent and viable organism. We can no longer take DNA to be arbitrary, for "if genes are just arbitrary DNA sequences, then most of them will have no more systematic relation to the phenotype than an arbitrary string of letters has to the meaning of a book" (Sterelny and Griffiths 1999, 79).

A phrasing of Hoffmeyer appears as follows: Suppose a living system arose from the primordial soup—or wherever it was. "We will have to ask: *Who was the subject to whom the differences worked on by such a system should make a difference?*" If one admits that living systems are information processing entities, then the only possible answer to this question is that the system itself is the subject. Therefore a living system must "exist" for itself. For a system to be living, it must create itself, that is, it must contain the distinctions necessary for its own identification as *a system*. Self-reference is the fundament through which life evolves, as it is the most basic requirement. But what is the basis of this self-reference, and thus the basis of life? The central feature of living systems, which allows for self-reference, and thus the ability to select and respond to differences in their surroundings, is *code-duality*, that is, the ability of a system to represent itself in two different codes, one digital and one analog. The chain of events that sets life apart from nonlife needs at least two codes: one code for action (behavior) and one code for memory—the very first of these codes necessarily must be analog, and the second (a code for memory) very probably must be digital (Hoffmeyer 2008, 81).

He states that the first set of these codes necessarily must be analog in order to embrace any increase in intensity of aspects of connectedness in living systems, while the second set of codes must be digital to continue to "remember" in the lengthy periods of evolution. The analogically coded signs "correspond to the myriads of topologically organized indexical and iconic [the terms are those of C. S. Peirce] semiotic processes in cells and organisms which incessantly coordinate body parts and their relation to environment. In so doing an analogically coded sign also is responsible for the interpretation and execution of the genomic instructions" (Hoffmeyer 2008, 81).

Digital coding, on the other hand, is "the unique mode of survival which is open to living creatures by reason of their mortality. We say that they survive semiotically inasmuch as they bequeath self-referential messages to the next generation." Digital codes ensure some continuity between the past and the future, since all such codes for memory must be immune to any fleeting transient variation.

In addition, Hoffmeyer argues, the system must be able to construct a description of itself. This description furthermore must stay inactive in—or at least protected from—the life-process of the system, or else the description will change—and ultimately die with the system. In other words, the function of this description-to-coding process is to assure the identity of the system through time: in all known living systems, this description-to-coding is made in the digital code of DNA (or RNA) and is eventually contributed to the germ cells where DNA code creates for all organisms their own forms of self-description.

If the fundamental condition of self-reference clearly depends on some kind of re-description, where does this occur? Here Hoffmeyer draws from Bateson: The processes of punctuation, which biology clearly recognizes and which is fundamental to digital form, are not inherent in the molecule, but are always conferred through preparing a contextual setting for interpretation during the course of evolution. Biologists should adopt the Bateson idea that bioinformation be considered reflexively, as a *difference that makes a difference through perception*. As Hoffmeyer argues, the difference that always makes a difference is to *somebody* who perceives (Hoffmeyer 2008, 30). The presence of myriad "somebodies" able to perceive indicates the immanence of both subject and a propositional order of signification all the way through the animal kingdom.

Next, Hoffmeyer draws from C. S. Peirce: it is the nature of a code to set constraints in organization, to "point outside of its own mode of existence—from the continuous to the discontinuous message, from the physical and therefore law bound message to the more free message. And back again in an unending chain. . . ." This sets life apart from nonlife, that is, it displays the unending chain of responses to selected differences. The *wonder of the code* is not stemmed by its enclosure into one (the physical) or the other (connective-interpretative) state space or life-world. Rather, life becomes a semiotic process carried forward in time by the historical lineages in which organisms are embedded—in their interaction with changing environments. Another quote may help explication:

> Had it not been for digital coding there would have been no stable access to the temporal world—ie., the unidirectional continuum of pasts and futures—and therefore there could be no true agency or communication. On the other hand, had it not been for the analog codes there could be no true interaction with the world, no other-reference, and no preferences. . . . Organismic "context space" expands at an accelerating rate in proportion to the increase in the semiotic sophistication of species; for, simply put, there are so many more different ways to be smart than there are different ways to be simple (and this may be the reason why the speciation rate among mammals is five times higher than the speciation rate among lower vertebrates). (Hoffmeyer 2008, 89–90)

Robert Rosen's notion of "scaffolding" then enters into Hoffmeyer's discussion of the formation of new functional genes (see Introduction). He refers to *semiogenic scaffolding* as the means by which a new functional gene occurs. He describes it as the lucky conjunction of two events, the first being the means through which an already existing, but nonfunctional gene, might acquire new meaning through reintegration into a functional or transcribed part of the genome, and the second being the conjunction of this event with the gene product hitting an unfilled gap in the semiotic needs of the cell or the embryo. "In this way, a new gene becomes a scaffolding mechanism, supporting a new kind of interaction, imbuing some kind of advantage upon its bearer. . . . For by entering the realm of digitality, the gene's new semiotic functionality becomes available not only to the cells of the organism carrying it, but also to future generations" (Hoffmeyer 1991, 138–139). In other words, conservation through time ensures digitality in the life sphere emerges from the contextual and therefore the more subjective experience in the system. Thus horizontal gene transfer permits a sharing of gene functionality with unrelated organisms and through this sharing process the observer is able to undertake some analysis of the "objectifying" of functions.

SKINS AND MEMBRANES

Hoffmeyer makes the startling claim that our membranes, body membranes such as human skin, are in charge of the administration of our life processes and "are as much in control [of] our life processes" as DNA. He takes up a strong postgenomic position and makes that which used to be called 'functional boundaries' with their thresholds, their surfaces and their changing processes of inclusion and exclusion, a central feature of his discourse. Membranes and their scaffolding constitute a prime locus for biosemiotic investigation. As he wrote me in a letter,

> The skin is only one among many surfaces of the body. Inside the skin, we find new surfaces around organs and tissues and eventually we come to the surfaces of single cells. The area occupied by the plasma membranes around our cells already comprises some 300,000 square meters. Inside individual cells, we immediately encounter new areas of membranes enveloping myriads of subcellular organelles, mitochondria, golgi membranes, endoplasmic reticulum, lysozomes, etc. All in all, a human body therefore consists of many square kilometers of membranes, and everything which goes on inside the body is more or less connected to proteins that are topologically ordered in relation to our extended membrane areas. Our membranes, rather than our DNA, are in charge of the administration of our life processes. And since the membranes logically impose a fundamental

asymmetry between outside and inside, membranes constitute the very locus of biosemiotic activity. That is to say, there is a continuous translation of externally occurring events into meaningful responses, or interpretants, at the inside of the membrane.

He suggests in his major publication on biosemiotics that differences across membranes express the basic asymmetry between organism and environment:

> A new understanding of the origin of life suggests itself. For whereas traditionally the origin of life question has been posed in chemical terms i.e., as the problem of accounting for the establishment of complex chemistry of proteins or nucleic acids (RNA or DNA) the semiotic approach sees the origin of life as the formation of a basic asymmetry between organism and environment. To ask about the origin of life is thus to ask about the origin of the environment, and this is fundamentally a semiotic problem of relating differences across membranes. (Hoffmeyer 2008, 17–32).

Since modulation in biological organization is always a result of multiple signaling processes, contextual selection must occur, and contextual cues bias genomic activity. The cellular (or organismic) system interprets inherited patterns inscribed in the DNA molecules according to the contextual situation in which the cell or the organism finds itself. All living systems, even down at the molecular level, have to resolve the fact that the information to which they respond is not always in the form of a discrete script, nor is it singular, nor is its response as inert as "on" and "off." Instead, living systems are sets of nested surfaces inside surfaces, canalizing an incessant stream of signs between exteriors and interiors. The failures of the Human Genome Project, discussed in the previous chapter, bolstered the supposition that wherever cellular membranes exist, contextual recognition is fundamental to the canalization of this continual stream of signals and signs.

Hoffmeyer refers to the possibility of cells being able to interpret meaningfully any message that may cross, either inwardly or outwardly, the membranes enclosing living systems. Another biosemiotician has suggested that interpretation in the context of the cell does not necessarily imply self-awareness or an elaborate knowledge of the contextual setting, "but simply the ability to sense the exchange of messages through the exposed interfaces, as signs satisfying their need for completeness. Any environmental sensing that proved capable of fulfilling their metabolic requirements would then be interpreted as a meaningful sign matching their expected survival needs" (Giorgi 2012, 15).

BEING-IN-RELATION

Many in biosemiotics, including Jesper Hoffmeyer, have found C. S. Peirce's ontology of signs a more enabling means to discuss signhood than Bateson, because they favor an ontological approach to biology rather than an epistemological one. For these adherents, Peirce's semiosis is always an ontology of "being-in-relation," rather than the usual portrayal of the ontological "being" per se. The triadic aspect of signification that Peirce put forward indicates that anything that is a sign is a sign—not as itself, but a sign in some relation or other. Thus, any coding of signhood yields a relational process, encasing the experience of life into an ongoing semiotic (reflexive) process relating to signs.

What resonates with Hoffmeyer is the sense of autonomy that the idea of being-in-relation brings to ontological argument for it infers that the organism always interprets as participant and therefore as subject in semiosis. Being-in-relation carries forward propositional notions of intelligence for example, the Peircian notion of "habit," which is both immediately expressible as "habit" by observers; yet, it is also evidently identifiable by participants, and can generate among participants propositions about future activity toward the "other." Habit, therefore, enters into the participants' ongoing relation with the "other." The derivation of selective behaviors, habit on the part of organisms, then leads to differences in sexual reproduction, and all of these differences seem to be tied to the meaning of relationships that organisms have within their eco-niche.

Semiosis requires proprioceptive sensing (the ability to sense or perceive stimuli arising within the body) to account for sign interpretation. Investigation of proprioception shows how animals incessantly sense their own movements and how this ability to sense generates a perceptual mapping that subsequently guides situated movement (we have seen with regard to human perception-motor theory in Chapter 4, where somatic processes are themselves subject to constraint from the psychological level). Biosemiotics therefore investigates, among other things, the recognition by living organisms of contexts in which proprioceptive events occur. Proprioception involves body/mind as unity and includes surround or environment in that unity. Hoffmeyer argues that organisms create their own umwelt, or ecological niche, in which "the *umwelt* is the representation of the surrounding world within the creature." Thus: "The *umwelt* could also be said to be the creature's way of opening up to the world around it, in that it allows selected aspects of that world to slip through in the form of signs. . . . The specific character of its *umwelt* allows the creature to become part of the semiotic network found in that particular ecosystem" (Hoffmeyer 1996a, 58). An ecological niche is a *semiotic niche,* where sets of environmental cues (visual, acoustic, chemical, etc.) must be correctly interpreted for the individual, or species, to survive.

A sign proposes something. Thus, a "habit" is the source of many signs, the sign of a wounded leg on an infant chick, or a broken wing. Distress cries on the water will yield several propositions about a particular sign (interpretant) before it becomes selected and related to interaction, all the way from a "true" event requiring immediate attention to deliberately misleading events, as in the case of a Canadian loon's lengthy performance in the water of being hurt with its wing supposedly broken, in order to detract predators from its new-born chicks.[2]

One of Hoffmeyer's justifications for highlighting "habit" is that he believes that "habit" is one of the most identifiable features of living systems, and that it distinguishes creatures that have habits from nonliving systems. Habits are one of a living creature's marks of their historicity, as opposed to the more spurious mechanistic notion of "instinct." By this Hoffmeyer means that what is passed on to the future is not just a trace from the past, as in the mechanism of instinct, but a kind of interpretation of the past in the context of the present. Or, in broader terms, living organisms undertake activity through their involvement in their eco-niche, which results in their developing "habit," which, in turn, enables the evolutionary incorporation of the present "fittedness" into the future.

Natural systems exhibit profuse examples of primary, discriminatory skills, or habits that have strong relations to categories of perception. When elaborated in evolution, these habits lead to communicative and eventually, in the case of humans, linguistic recognition, of "self" and "other." Thus, "the key to scientific understanding of the mental is embodied existence [habits and their influence on perceptive and cognitive anticipation] and [is] not the fictitious idea of disembodied symbolic organization which appeals so strongly to the arithmetic formulations of science" (Hoffmeyer 1996b).

This leads him to his concept of *semethic interaction*: "Whenever a regular behavior or habit of an individual or species is interpreted as a sign by some other individuals (conspecifics or alter-specifics) and is reacted upon through the release of yet other regular behavior or habits, we have a case of *semethic interaction* (from the Greek *semeion* = sign plus *ethos* = habit)" (Hoffmeyer 2008, 189). He further suggests, in the same vein, that organic evolution itself shows a gradual increase in *semiotic freedom*, or depth of meaning that an individual or species is capable of communicating through semethic interactions.

C. S. PEIRCE AND ABDUCTION

We come now to an idea that Bateson borrowed from Peirce, that is, Peirce's notion of "abduction," which can be used to account for the process that provides the all-important transition from percept to trial-and-error judgments in creatura. Most people know and understand the way in which inferences can be drawn from induction and deduction, the one belonging to the methods of science and the other belonging to the methods of rhetoric and logic. The methodologies of induction and deduction are part of school curricula. People are less likely to have heard of "abduction" as a logical process. The term "abduction" relates to "best guess" or "conjecture" or even "feasible solution" to some form of proposition. How do these derive from perception? We have already come across "conjecture" or "guess-work" in relation to perception in Chapter 4 and seen the arguments of reverse hierarchy theory, in which explicit vision is achieved through a sort of "guessing" of categories from obscure information received. And Chapter 5 discusses Bateson's use of logical types, which are another form of propositional order that accompanies "hunches" about communicative sequences in human interactions.

Peirce regarded deduction and induction as two possible logics of semiosis in the human sphere, but to deal with a generalized relation of sense and perception to feeling (or Firstness), he proposed another means of inference-making, which he called abduction. Abduction is conjecture and it changes the rules about inference-making, for abduction allows for analogy; it does not require the sequence of steps that are part of induction or deduction nor their strict methodologies to ensure verification. In Peirce's methodology, the steps are three in number: Firstness, Secondness, and Thirdness; induction and deduction can only derive from Thirdness. Nevertheless, there is a sort of method, abduction, attached to Firstness.

As Merrell describes this contribution:

> Abduction, in contrast to the other two inferential processes [induction and deduction], is a tentative conjecture that if a certain situation happens to be the case, then some imaginary set of consequences might possibly follow . . . abduction chiefly entails sign possibilities of Firstness, [while] induction is primarily a matter of signs actualized or Secondness, and deduction posits that something should, or would or could be the case given a set of conditions; hence deduction is primarily a matter of Thirdness. (2003, 199)

A generalized relation of Firstness, or feeling, or a state of possibilities, to use Peirce's terminology, could be held to occur in all animate existence. Abduction in a generalized sense is the means by which an initial focus occurs, is perceived, and in which abduction affirms something about the perception, which then enables the initial interpretative hypothesis, a vague feeling, to become a perceptual judgment. For Bateson, the benefit of Peirce's argument about abduction is enormous because conjecture of abduction makes intelligible how an animal can draw information from any aspect of perceived redundancy (pattern) as information. This is also a proposition about the way in which creatura (nonhuman animals) demonstrate awareness. They do not have to have the capacity to work with predicate logic: they can abduct through experience, undertake analogic comparison and engage in some aspect of learning. In short, abduction is a process that exposes the fallacies of what Millikan called "meaning rationalism."

Abduction is a transform of coding from perception to propositional order. In the case of human beings, Peirce said, "there is abduction between any stated hypothesis and the perceptual content of a sensory process it represents." Thus, "What is described of what I see is different from what I see. What I describe is a proposition, a sentence, a fact; but what I perceive is not a proposition but only made intelligible by means of a statement of fact. The statement is abstract; but what I see is concrete" (Pape, 1999, quoting Peirce 1901, MS 692). Abduction therefore draws from concrete perception a conjecture about a proposition. That is: "what is described of what I [consciously] see is different from what I see." The transform is abduction. In the case of human beings, abduction has a syllogistic structure, but one that is distinct from induction. The classic example of abduction is about beans in a bag, as follows: "These beans are from this bag" (perception), followed by "these beans are white," which is based on the relation perception to an observable result, followed by "all beans from this bag are white." The guess that is assumed is a conjecture or an a priori assumption that is true of experience. This indicates a conjectural inference. By this means, "habit counts."

The "proof" of abduction accepts experience of behavior in anticipation of the events, as they are perceived. It is this aspect of abductive inference or conjecture that can be extended to the nonhuman animal realm. Percepts can be in some respects analogues of logic because their propositional form resists reduction to pure subjectivity. Thus, the process of mapping descriptions of pattern on to a tautology does not require any claims of exact isomorphic representation of the outside world, those abstract configurations required for quantitative analysis of things—measurements or other entities of "pleromatic" explanation. The process must only claim to be internally consistent, and thereby ascribe to the external world a consistency that will somehow resemble that which obtains within the tautology (Bateson 1975, unpublished).

Any perceptive mark about situation A can also apply to B and through an analogy enables the mapping of the pattern to mediate meaning.

Bateson takes these Peircian ideas about abduction and fits them it into his epistemology of information. The appearance of pattern within biological systems enables a mapping of the appearance of these patterns or phenomena on to one or more "tautologies," and preferably one "tautology." By this Bateson means that an explanation derived from one mapping does not need to assert the truth of its premises (as in the classic case of logical positivism), but can embrace a less strict definition, a definition that follows from expectation of prior experience. All abductions relate "sets of differences [that] are cases under the same rule." This type of explanation is not determinate in the strict inductive or deductive logical manner, but it can be sufficiently rigorous to be treated seriously. Thus, he wrote, there is an important overlap in the two concepts of tautology and metaphor for the cluster of phenomena presented in the metaphor, and the cluster embodied in the referent are both—by abduction—cases under the same tautology (Bateson 1975, unpublished).

Though schoolchildren are reprimanded for writing in a tautological manner, Bateson described tautological inference as the major means through which organisms in the living world are able to discriminate "this" from "that." His understanding of the term "tautology" is that an explanation does not need to assert the truth of its premises (as in the classic case of logical positivism) in relation to facts. Instead, its claim to be taken seriously refers to the rigor with which a judgment is made about pattern and its composition. "Mind and nature have a necessary unity" is a tautology. So is the statement that "the unity of an ecosystem is to be explained by all possible connections within it." A tautology must be internally consistent, and must have a logical consistency that somehow resembles that which obtains in the external world or it will not be believed. The rigor of tautology is aided by metaphor and other similar imaginative forms. In the case of metaphor, the cluster of phenomena presented in the metaphor and the cluster embodied in the referent that is being carefully described are both, by abduction, cases under the same tautology. Both cases embody same or similar relations in a topological form, and aesthetics, which we will come to next, is also a prime example of this.

A tautology can remain "true" because the mapping of descriptions of pattern on to a tautology does not require strict claims of representation defined by "fact." Nevertheless, there are important differences in Bateson and Peirce's way of constructing abduction. Peirce argues that the triadic relations of object-sign-interpretant begin with the organism *tracing similarities*, matching similarities of feeling with a ground (already itself a sign of some sort through prior experience) according to his dynamic constructive scheme. Bateson's starting point is the concept of difference in perception in which the difference that makes a difference becomes a primary motivation. Bateson argues that, initially, perception enables distinction to be made between pattern and noise. Perception requires transforms of coding of light diffused across the retina. What can be learned and examined from percepts lies in these transforms. Initial coding may be entirely unconscious, but it is possible to become aware of transforms in the process of recoding, or grasping contexts. Recognition of similarity and resemblances then occurs with the ordering of contexts. Repetitions of transforms of differences cross boundaries between other recursive systems—perceptual, cognitive, and cultural—and build up a variety of contexts, including, as we shall see in the next chapter, the metapattern of aesthetics.

In other references to abduction, Bateson argues that the subject matter of abduction is "precisely the relations between the components of the aggregate of the phenomena." These are "the ideas" of their organization. Are the resemblances older than the differences? He takes the example of a plant and a crab. They could be said to be resemblances between the differences, or even "resemblances between the differences between the resemblances between the differences." He notes that one half of the resemblances lie in the chromosomes of the two, plant and crab, and the other half lie in the microscopic resemblances of what is called the phenotype (Bateson 1973, unpublished).

We can note that Hoffmeyer's own stance is abductive when he comes to discuss semiosis in cellular contexts and is an unfolding of "the pattern which connects" through rigorous tautological comparison. When Hoffmeyer discusses human subjectivity in relation to cellular subjectivity, he does not have much empirical information to go on regarding the latter, but he can make a well-informed guess as to the comparative aspects of how the two could relate to each other. After he chooses a comparison based on abduction, there are a common means or way of deriving "meanings." This begins with a recognition of the difference, and the sifting of the evidence then depends on defining the limits of comparison in the tautology.

THE MEDIATION OF AGENCY

Biosemiotics maintains that life and sign functions are coextensive. Any organism able to perceive "double differentiation" between expression and content of expression (i.e., is able to create a context) is certainly able to recognize "sign" and exhibits processes of signification. In fact, biosemiotics maintains that wherever there is a recognizable distinction that enables the participant to identify a context, there one can begin to speak about semiosis. Equally, if the participant organism or cell is able to make a distinction between an identifiable "self" and an identifiable "other," there too can one begin to speak about semiosis. These, when elaborated in evolution, lead to communicative and eventually to linguistic recognition of "self" and "other."

Other Peircian interpreters, such as those in cognitive semiotics, disagree. They argue that the full features of the Peircian analysis of mediation of mind require conditions of consciousness, and these conditions appear late in evolution and are not apparent in earlier evolutionary forms. The notion of "meaning" in nonhuman animal relations cannot therefore be properly be correlated to "signs." While a word or a picture in the human semiotic context is a sign—in the common sense of that term—hormones, antigens, and antibodies are "signs" only in a metaphorical sense. Thus, the biosemiotics perspective of "the systematic study of meaning" dips too much in the realm of metaphor.

Cognitive semioticians claim that biosemiotics has projected common sense resemblance between human relational activity and relational activity in the biochemical sphere much too far—far enough that biosemiotics' commonsense analogy cannot be pushed much further into scientific methodology. The cognitivists point out that Peirce's phase of Thirdness is strongly associated with the sort of conditions that lie within a human domain of linguistic communication, manipulation of symbolism and institutional organization (law). In other words, if there is to be a Peircian application of semiosis, this can arise only in a strictly defined anthroposemiosis where symbolic mediation is consciously construed. And here, in the cognitive semiotic view, the appearance of symbolic signs (symbols) is coextensive with language (Zlatev 2009).

The cognitivists are, of course, correct but their conclusions about the validity or otherwise of biosemiotics ignore the obvious point that the Peircian scheme is at fault for having so little interest in the world of creatura. As the cognitivists themselves point out, it is only with a fully fledged Peircian notion of Thirdness, based on a conventional ground of signs as symbols, through emblems, words, language, or grammatical constructions, that "signs" in the proper Peircian sense of the term, appear. They say conditions of Thirdness in the Peircian scheme occurred long after humans and protohumans detached themselves from the other primates. Quite true. Even if primates see similarities between the members of their own group and may recognize differences among them—since they can recognize individuals—they do not see one member as a sign of another, they say. This could be true, but since the Thirdness of Peircian semiotics is language-dominant, and rational-learning dominant, it is unlikely that their Peircian criticisms can say very much either way.

The first forty years of Peirce's writing, especially his "transcendentalist perspective," applies only to human beings, the internal aspects of mind, and its basis in logic and truth conditions. There is little room for biosemiotics here. Those who study Peirce notice a change about 1907, some forty years after his original derivation of semiotics, when he began to think less about the logical propositions of "sign" and extending his arguments to communicative signs. At this point he began to deal with signs that are uttered and interpreted. It is here that he introduces the concept of a dynamic interpretant and turns his logical categories of Firstness, Secondness, and Thirdness into phenomenological categories of thought. Nevertheless, Peirce's post-1907 processes of sign-interpretation remain focused on self-controlled habit formation and a rational learning process. In other words, Peirce continues to use the concept of sign as a representation about something that rational or scientific inquiry is supposed to reveal. After a thorough review of whether Peirce's thematics enables an easy transition to biosemiotics, Tommi Vehkavaara, himself a member of the founding biosemiotics group states that Peirce's continuing requirement for self-controlled conduct "restricts the possibilities to apply Peirce semiotic concepts in biosemiotics severely" (2005, 291–292). There should be no demand for biosemiotic agents (creatura) to offer conscious explanations for their action. As Ruth Millikan might say, Peirce's mediation of mind, indeed his whole concept of Thirdness, remains too encased in the conventions of meaning rationalism.

Like Bateson and Millikan, Hoffmeyer rejects language dominant theories. Scientific views become anthropocentric when human language becomes the distinguishing characteristic of species performance, and by extension become an ultimate criterion for any study of meaning and interpretation in living systems. For that reason, Hoffmeyer also rejects what he finds in Saussure's interpretation of linguistic meaning, because Saussure offers only a dyadic interpretation of linguistic signification. Peirce's mode of triadic explanation is at least efficacious (Emmeche and Hoffmeyer 1991) in respect of biological habituation: "habits" give rise to expectancies and therefore also to perceptual and cognitive anticipation of "others" in self/other relations. They form the basis of self/other perception, which, in turn, is a basic characteristic of mind and marks a process in evolutionary development. But it is no wonder that Millikan began her enquiry with the notion of a natural signs, rather than Peircian signs and specifically limits her study to interactive communication in local systems.

As for Bateson, he recognized the merit of Peirce's scheme being monistic, rather than dualistic, which he took as a great advance in thought about the relation between mind and nature; but he was unhappy with that element of

Peirce's triadic schema. Bateson disregarded Peirce's category of Secondness, which drew upon material or "factual" dimensions of intentionality and embodied these processes in his monistic approach. Instead, Bateson substituted "difference" and "context" and context markers. As shown in Chapter 7, he also developed his own triad with respect to "difference." To repeat: Which difference makes a difference in any ecosystem is "a threesome business," he wrote, the first term of the triad is circular causation at several levels, the generation of form per se. The second term is that of co-learning in adaptation, and the third term is that of genetics, which is a memory system, and a much slower temporal variable than the other two. Each draws upon the other recursively as relations of the other.

We may reflect that Peirce's whole scheme of Thirdness, as the mediation of mind, is only hesitantly recursive. Moreover, his depiction of mind to matter was relative simple, formed perhaps from his attachment to Unitarianism; it was of mind and matter being two sides of the same coin. It expressed no subtleties such as those contained the Gaia hypothesis that living systems mediate material conditions on earth. Nor did it express Bateson's notion that mediation might switch from virtuous circles to vicious circles and threaten the stability of living systems.

PROTOSEMIOSIS

Those in the biosemiotics group critical of Peirce point out that his ontological approach lacks *agency*, and that this lacunae is particularly important as one begins to examine the very borders of life, such as research on bacteria and viruses and simple cells (Fernandez 2010, 3). Simple agents are not capable of associating with "objectives," which of course, is the precondition of Secondness in semiosis. Sharov is one person urging biosemiotics to forgo standard Peircian approaches to semiotics in the case of microorganisms and instead build an approach to protosemiosis instead.

There are others in agreement, arguing that most "organisms can't change their objective ontogenetically for they are tied to certain objectives given by perception-action cycles." At this level of enquiry, biosemiotics is dealing with simple shape recognition, in which animal reacts in a general way if and when neutral objects, not related to immediate survival, begin to appear in organism's umwelt. Thus an observer will have to begin "with holophrastic or general signs," as Sternjfelt put it in an online discussion, before proceeding toward specific sign use of the sort that Peirce was proposing.

For much of the last fifty years, biologists outside biosemiotics have thought of viruses as nonliving forms, that is to say, microforms that failed to exhibit crucial characteristics of living microorganisms. Viruses were described as "sorts of cells" that had been derived from the cellular world through gradual loss of cellular functions, forcing them into parasitism in the cell. Therefore, they were declared to be "dead" microforms, and were instead referred to as "elements" or particles. Today much of that initial list of absent properties of life, first put forward by Andre Lwoff in 1957, has been proven incorrect (Claverie and Abergel 2012, 194–196).

Researchers now recognize a type of swarm intelligence existing in microorganisms. The swarm intelligence phenomenon, with regard to consortia of bacteria, is visually captured on YouTube. Eshel Ben-Jacob, who conducts research on swarm intelligence, offers the following perspective on bacteria. We use antibiotics indiscriminately, Ben-Jacob says, unaware of bacteria's social intelligence, which allows them to learn from experience to solve new problems and then share their newly acquired skills, and so on. As a result, bacteria developed multiple drug resistance, and we cannot invent new drugs fast enough to beat them. To change this threat to our health, we must realize they have social intelligence.

Swarm intelligence also exists among viruses, although this is more difficult to discern, since viruses are of several orders of magnitude less in size than living cells and are smaller than bacteria. But Günther Witzany, a member of the biosemiotics group, argues that the prevailing idea of viruses needs to be altered. The view among the public is that they are destructive agencies. In some ways this is undoubtedly true. There are destructive viruses or virions. But it turns out that viruses have a double form and that they also aid in the perpetuation of life. There are viruses that persist in the cell; moreover, their persistence in the genome of the cell is of mutual benefit over time. Both cell and virus are symbiotic in the case of persistent viruses. On the one hand, the virus cannot replicate without association with the cell, which means that the natural habitat of viruses is that of cellular genomes; on the other hand, persistent viruses, or nondestructive viruses, actually *filter* those infecting agents by editing the genetic code. They are indeed double agents with double meanings for the sustainability and advancement of life.

The twin life strategies of viruses, both of them completely different, can be found especially among living organisms in the sea, where they are both a major source of mortality and persistent viruses that broaden life chances by giving immunity to organisms. This new understanding of the double activity of viruses has led to a change in the whole concept of the genome among those studying life at the edge of living systems. Previously, the genome was thought to be a storage device, but now the genome appears to these researchers to be a species-specific ecological niche in which persistent viruses have viable host-to-virus relations in that niche. Most important, viruses are part of the dynamics of an ecological system called the genome.

In Günther Witzany's view, persistent viruses are also the agents of genetic invention, that is to say, they are masters of the identifying sequence-specific loci of genetic text and, if need be, editing the genomic text in order to enable recombination, repair and regulation in cellular life. Witzany believes that viruses are competent agents able to undertake genome editing. Viruses are able to change many regulatory elements vital for the expression of new patterns or, alternatively, the silencing of the genes. In this agency-based perspective, every coordination process between cells, tissues, and organisms depends on signs between signaling agents. Agents are competent to use signs by generating them, receiving them, and interpreting them, while signs transformed into signals trigger behavioral responses (Witzany 2012, p. x).

Witzany's analysis amounts to a radically different prospectus from that of any view that stresses viral agency as random assemblies of physical entities of nucleic acid sequences undergoing mutation. Instead, he sees the agency of viruses as coevolutionary partnering with a host organism, in which viruses acting through networked association become their own consortia. The tracking of consortia, or virus networks, during the last twenty years confirms the ability of viruses to generate populations.

> The consortia is able to make errors which are then overcome by the consortia acting as a unit. In other words, there is ongoing feedback taking place. This does not mean that all the interactions within a quasi-species population of viruses are complementary or supportive. Some interactions will always involve lethal interferences to the cell. Moreover, distinct, quasi-species populations do indeed compete with and exclude each other. Nevertheless they can develop internal coherence and identity that expresses a relative fitness that enables persistent viruses to exist in host-virus symbiosis for long periods of time. (Villareal 2012, 117)

The fact that consortia or quasi-species act as networks, and not as an autonomous individual virus, means that the modes of activity of the whole consortia together ramify throughout its own network. Thus, the whole population within the network must be familiar with the code within the genome, Witzany maintains, in order to act upon it. Indeed the consortia would not be able to undertake the necessary editing procedures of the genome in order to maintain both network and cell fitness if they were not familiar with coding text. When acting on the coding text, network agents do not act in a strict sequential manner on the syntax of molecular codes as is generally assumed in the Darwinian paradigm (Villareal 2012, 113).

In summary, the consortia phenomena, through a network of agents, reveals the hitherto unknown capacity of viruses, namely that viral agents are natural genome editors. Witzany points out that viruses show competence for code reading, deleting, and inserting in a context-dependent way, and their competency extends to their ability to develop contextual responses.

9

ECOLOGICAL AESTHETICS AS METAPATTERN

> I believe that much of early Freudian theory was upside down. At that
> time many thinkers regarded conscious reasoning as normal while the
> unconscious was regarded as mysterious, needing proof, and needing
> explanation. Repression was the explanation. . . . Today we think of
> consciousness as the mysterious, and of the computational methods of
> the unconscious, e.g. primary process, as continually active, necessary
> and all embracing. These considerations are especially relevant in any
> attempt to derive a theory of art or poetry.
> —GREGORY BATESON

Earlier accounts in mainstream science usually portrayed sensory perception
as a passive window through which the mind accepts sensations. These in turn
were selected and assembled somewhat like the pieces of a jigsaw puzzle as
the "sense-data" of perception. If perception is *not* passive, as Bateson always
maintained, and if sensations are in some ways created by the brain, the external data picked up by the visual or other senses can hardly be the sole data
for our perception of the object world. There is also imagination involved, and
memory.

The unconsciousness or nonconsciousness of the processes in our mental
states was never a problem for Bateson. He pointed out that there is total unconsciousness of the processes of perception, that is to say, our mental "machinery"—or what we know about it—provides only news of its products of
perception and no news about the processes of perception. At the same time,
he acknowledges that most people assume that they see what they look at. Perception gives us knowledge of the states of our body in relation to objects and
events in an environment, as a result, people discount the unconscious nature
of the processes of perception. Yet:

when we say that what we see, feel, taste, hear some external phenomenon, or even some internal event, a pain or muscular tension—our ordinary syntax for saying this is epistemologically confusing. What I see when I look at you is, in fact, my image of you . . . these images are seemingly projected out into the external world but . . . "the map is not the territory," and what I see is my map of a (partly hypothetical) territory out there. (1991a, 204)

At first he regarded this phenomenon as being a sort of biological deficit of our field of consciousness. The early work that he did on consciousness was strongly influenced by the cybernetic analogy, life as an ordered series of cybernetic circles in the form of bodily homeostasis. Thus, a cybernetic system, like the brain, could not register complete consciousness since its neurons reporting on events at any given point in the system could not, at the same time, report on "effects" of new neural loops formed by conscious interpretation within its whole reporting system of neurons (*Notebook* 11, December 1949). The cybernetic looping was always out of the joint of real time.

Bateson then turned toward elaborating a synthetic view of mind with respect to perception and purposive consciousness. His 1967 article on conscious purpose versus nature (2000, 432–446) expresses a semipermeable linkage between consciousness and the remainder of total mind. A certain limited amount of information about what is happening in this larger part of mind seems to be relayed to the screen of consciousness. Because the *whole* of mind is unable to report to the *part* of mind, consciousness receives a systematic sampling of the rest of mind; what gets to consciousness is selected. Thus, much of input may still be consciously scanned, but only after it has been processed by the totally unconscious process of perception. What was needed was an investigation of how particular thoughts fitting the particular parts of the context of consciousness completed the whole circuits of consciousness and unconsciousness, acting in a more complex manner than simply as "bridges" between the conscious and the unconscious. Until 1968, Bateson wrote of this partial blindness of biological consciousness as the main cause of our cultural pathology.

He changed his view about the biological underpinning of conscious purpose as a result of two conferences in 1968, both of them surrounding the question of conscious purpose and human adaptation (2000, 446–453). After the conferences, he came to the conclusion that decision making in very large systems changed *the ratio of the power* (in the sense of "presence") between consciousness and the *rest* of human mind (MC, 11/11/69, 733-2). It was not so much the biological cybernetic organization of mind that had created the overwhelming appeal to rationality in devising our conscious purposes and that was creating systemic distortion. Consciousness pertains to social knowledge and not to individual self-knowledge. Social knowledge is knowledge shared by the individual with others, or *"con-scious"* [*sic*] sharing.

Scientists in biology and neurophysiology remain stuck with an individualistic, neurophysiological approach to consciousness. They argue that we do in fact grasp our own purposes, and that there must be an explicit representation of each aspect of visual consciousness in an individual in order to explain how this occurs, with each representation becoming clearly coded (Crick and Koch, 1998). Formulating the "how" of neural coding will then provide definite characteristics of consciousness. Why does science argue in this manner? From Bateson's point of view, scientific epistemology considers it necessary for consciousness to be explained in a physical manner, as an outcome of physical interactions. If there are nonphysical attributes, such as "unconscious perception," which yields meanings other than the "truths" of coding, then a physical explanation will no longer have a unique materialist framework, nor will it justify scientists' beliefs in the conjunction of explanatory truths and "objective reality."

His own view was that if he was correct in assuming that consciousness is social, then science had to settle for limitations in its attempts to trace the neurophysiological capabilities of consciousness. For the self-purposiveness of consciousness, almost of necessity, is unlikely or unable to act in accord with any systemic setting all the time; the latter is too complex. In an ecological setting, for example, the whole setting or system may be consciously appraised, but individual purpose will instead follow the shortest logical or causal path to get quickly to what one individual may wish to do: cut down oak trees for immediate use. Thus, oak forests have virtually disappeared (Bateson 1980, 77).

He had held from his early days as a psychotherapist that no human individual has a secondary system of consciousness that is not mediated or accompanied by a primary system (MC, 01/02/49, 879–8a). In no animal does a cortex exist without the lower parts of a central nervous system. Likewise, there are no separate divisions within the brain dealing with "reality" of consciousness on the one hand" and affective feeling on the other. Thus, when it comes to purposive action sequences, it may be irrelevant to invoke criteria that distinguish consciousness from unconsciousness—except that when answering questions about our own purposes, a selected event or a picture is drawn, and this process limits finding our own self-conscious purposes. On the other hand, in any interaction sequence invoking "play," a person may see himself as "playing," while unconsciously he is "not playing" and vice versa—he may consciously think he is "not playing" while unconsciously thinking he is playing (*Notebook* 14, 1951).

Science may be correct in assuming that it is necessary to reveal codes of consciousness to gain an explicit representation of purpose, yet primary process did not simply place before consciousness a random assortment of instructions. Instead, it requires transforms of coding to join the links between experience of vision, processes in the cortex, and subjective states. He argued that it is not the codes that should be the central focus of enquiry, but the *transforms of coding* that render the intersection of multiple coding in all areas of the cortex meaningful. Initial coding is likely to be entirely unconscious.

By ignoring all this evidence about the importance of the unconscious in forming our perceptions, we lose the possibility, the wisdom, of grasping knowledge about ourselves as a whole. The whole, as he describes in *Angels Fear*, include differences ("distinctions" or "indications") derived from coded transforms that may be imaginary or may cross boundaries between unconscious self-reference and other recursive systems—perceptual and cognitive, cultural, and natural—as all of these build up varieties of contexts. Through the examination of these contexts, what we come to understand better are the processes of linking, relating, and connecting *the transforms* of information.

Aesthetics is one means by which we move to interfaces of these transforms of information in order to capture the pattern of transforms and so enlarge our consciousness. "I do not know the whole remedy," he wrote, "but consciousness can be a little enlarged through the arts, poetry, music and the like. And through natural history [as well]. All those sides of life which our industrial civilization tries to mock and push aside" (MC, 28/71968, 1056). Aesthetic responses that rely on the imaginary are a help for aesthetic appreciation at an interface "wherein lies the integration of aesthetics" (2000, 470), and an integration of the various levels of mental process lying in between.

To repeat one of Bateson's best-known quotations: "Aesthetic" means responsive to the "pattern which connects." "The pattern which connects is a meta-pattern. It is a pattern of patterns. It is that meta-pattern which defines the vast generalization that indeed it is patterns which connect" (2000, 11). Aesthetics enables perception of the recursive and holistic aspects of ecology; it gives human beings an ability to grasp an "algebra" of relations that underlies order in ecology, though any precise description of the underlying order of ecology is difficult to express. Aesthetics develops a metaperspective of holism partly derived from abduction in its many imaginary perceptions of beauty. Above all, it helps ensure a moral divide between creation, the sacredness of creation, and the destructive effects of nuclear war and biocide on lands and in oceans, seas, and atmosphere.

ART AND ANTHROPOLOGY

Bateson's aesthetic thinking in his younger days began with his research on William Blake. It was research that he never fully abandoned. Blake always remained an important prompt for his aesthetic ideas. His affinity for William Blake was partly a result of his father's purchase of a Blake painting, "Adam Exulting over Eve," which occupied pride of place in his boyhood home in Grantchester. One of his elder brothers, Martin, undertook a lengthy study of Blake's life as an artist, and his brother's unfortunate death prompted Bateson to undertake his own research about Blake's artistic merit. After his father's and mother's deaths, the painting was transferred to Gregory Bateson's possession. His Blakean interests in art did not carry over to his very first fieldwork study among the Baining, though prompted his interest in second fieldwork study of the Sulka peoples. The Iatmul, the focus of his third fieldwork, lived in the Sepik area of New Guinea, an area that was highly productive in its masks, figurines, and other ritual artifacts relating to totemism. He saw the totemic art of the Iatmul as being "decorative or "ornamental" relating to the Iatmul people's strong association with their totemic ancestors, but this did not lead him to any classification of Iatmul carvings as art per se.

Mead and Bateson's collection of Balinese art was a much more deliberate attempt to represent cultural achievement, and they anticipated that their collection of Balinese paintings would be central to their respective research. As we have seen, the events of war overtook both of them, and the artifacts were packed away, scarcely examined, until Hildred Geertz opened the boxes some fifty or so years later (Geertz 1994). The couple relied upon quick selection of the many photographs they had taken, a few of the artifacts they had collected such as the upside-down god, and the film work they had done to support observations about dance and trance in order to represent Balinese culture.

Added to all this was the cultural presuppositions of their audience in the United States. "Primitive art" drawn from societies around the world was not yet "art" in the Western imagination, unless that art was of societies considered to be important to the history of Western civilization. This public response marred Bateson's curation of Balinese culture at the Museum of Modern Art in 1943 and occurred again three years later, in 1946, with a major exhibition at the New York Museum of Modern Art of art in Oceania. There were more than four hundred pieces in this exhibit, stretching from Easter Island to the Marquesas, Fiji, and New Guinea, but the reviewer in *Time* noted that some visitors were shocked by "the obscene gods," crocodile gongs, and grinning masks. He remarked, "the more studious reminded themselves that art history begins with magic rites" (*Time* 1946). Bateson's own review for this important exhibit felt that the exhibit showed how much of culture is an act of participation in the rhythms in the human body, specifically in how much human culture is symbolically determined by the human reproductive cycle. The exhibit had also proposed a hypothesis about ethos, affective feeling, and the psychological place of artistic conventions in human life—all of this without specific reference to Freud. Bateson applauded this feature, since "the whole of Freudian theory is . . . so flexible that it is exceedingly difficult in practice to suggest any conceivable fact which would disprove any part of the theory" (Bateson 1946b, 122).

Bateson returned to an examination of Balinese art in a 1967 article, which stands as one of his major statements about art and aesthetics (Bateson 2000, 128–152). He picked out a depiction of a ceremonial burial complete with a Balinese cremation tower. A good art style, Bateson comments, is one where the individual artist imparts "a correction in the direction of systematic wisdom of natural oversimplifying or polarizing tendencies [in either] a culture or individuals." So the artist may attempt to balance or even oppose certain valuations or polarities in spatial perceptions, and these would appear by the artist's reducing or reversing the gradients in form. As an example, Bateson observes that the artist composed Balinese imagery of a turbulent lower field in his painting (e.g., those of the upside-down gods), which was then contrasted to his serene upper field. That "contrasts with the western tendency towards the reverse," for the Western painter would likely choose a rotated gradient in which the turbulence of spirits would be painted in the upper (heavenly) field.

He noted that the Balinese painting also brought out very different emphases from that which would be captured through use of a camera lens when taking photos of the same ceremony. He had taken many photos of the building of the cremation towers, of their bearers who carried them. Interestingly, he goes on to suggest, without quoting any authority on the issue, that cultural aspects of artistic perception "may serve to modify inherent gradients in perceptual motor systems and limbic/brainstem tuning arrived at through ecological and cultural interactions during an individual's development."

Bateson's article on style and grace in Bali takes a hard line against the relevance of Freudian ideas to the artistic imagination. "Mind is not like an iceberg," Bateson wrote, countering one of Freud's better known aphorisms; nor is that which is seen at the tip of the iceberg of the same form as that which lies beneath the berg, as is commonly supposed. Classic Freudian theory assumed that artistic productions, like dreams, are drawn from the unconscious and filled with thoughts that could be made conscious, but that repression and/or dream work had distorted. Thus, the Freudian view of the unconscious was that of a cellar, or a cupboard, to which fearful and painful memories are consigned by a process of repression, and which could be translated into rational language, but only with difficulty.

Freud also maintained that when dream content is unacceptable to conscious thought, it becomes translated into the metaphoric idiom of primary process to avoid waking the dreamer. Bateson retorted that such a view had its limitations. The idea of the unconscious as "the cellar of mind" might be satisfactory to describe the effects of psychological repression, but Freud's view postulated that there is a vast range for *consciousness* in mental process and that only a relatively small range for the *unconscious* in mental process. The reverse is the case. Like many thinkers in the Western tradition, Freud regarded conscious reasoning as normal, while the unconscious was regarded as mysterious, as needing proof and explanation. At the same time, Freud also assumed that the use of logic during consciousness was repeated in the idioms of the unconscious. Freudian theory was "upside down" in these respects: The realm underneath the iceberg is not of the same logical type as that smaller portion controlled by the observer.

THE AESTHETIC IMAGINARY

As an alternative to these Freudian views of the relation of art to aesthetics, Bateson turned to R. G. Collingwood. Like Peirce, Collingwood finds that the relation between the act of feeling and what we think is different in kind: those relations between what we feel and what we think become transformed in the pathways between them. But unlike Peirce, "The experience of feeling is a perpetual flux in which nothing remains the same and what we take for permanence or recurrence is not sameness of feeling at different times, but only a greater or lesser degree of resemblance between different feelings." Artists do not imitate, nor does art "represent" in any simple manner, for any given representation is always a means to an end. "A representation may be a work of art, but what makes it a representation is one thing, but what makes it a work of art is another" (Collingwood 1979, 158–161).

Collingwood proposes that perception and imagination lie in the gap between feeling and intellect. Perceptions may be fleeting phenomena that derive from the senses of touch, smell, sight, and so on, but the point at issue is that sense alone—sense as data—is never sufficient for artistic expression. Collingwood goes on to argue that perception and emotion are not rendered through concepts derived from abstracted immediacy. Instead of a direct transfer of sensation through consciousness to thought, or instead of thinking of aesthetics as a direct representation of external beauty, there must be additional sequencing, reflective moments of experience, which can be described as the return of the mind upon itself. It is feedback of the reflective mind upon itself that renders conscious attention on the experience (Rubinoff, quoting Collingwood, 1970, 55). Thus, perceptions are simply a primary experience, a primary occurrence that conscious attention then seeks to render explicit from what is implicit in the experience.

As Collingwood argues, artistic expression does not begin with emotional stimulation that proceeds immediately to an artist's representation of a perceived form—rather, the attempt to express such a form is always a process of becoming conscious of emotion (Kemp 2012). As conscious awareness of the emotion goes on, the forms of expression become a feature of artistic utterance. Clarity results when the focus of mind on the form of expression either intensifies that which is felt in the first place, or attenuates that feeling. In effect, any work of art develops by means of feedback between painting and community experience over time.

A third term enters at this point, named *the imaginary*. Though it has been commonplace for several centuries in Western thought to confuse the imaginary with false sensation, or with the illusions of "make believe," nothing could be further from the mark. Imagination is a form that emotion or feeling can take when transformed by the activity of consciousness. Imagination lies partly between sensa and thought; it is not sensa directly, but is sensa with respect to which the interpretative work of thought has been done well, or done ill, or left undone: "A real sensum means a sensum correctly interpreted; an illusory sensum, one falsely interpreted. And an imaginary sensum means one that has not been interpreted at all; either because we have tried to interpret it and have failed, or because we have not tried" (Collingwood 1979, 194). One commentator termed Collingwood's discussion of the imaginary as introducing "ideated analogues." It is a forceful expression, for as Collingwood argues, the imaginary is very close to standard notions of "an idea":

> There must be . . . a form of experience other than sensation, but directly related to it; so closely as to be easily mistaken for it, but different in that colour, sounds, and so on which in this experience we "perceive" are retained in some way or other before the mind, anticipated, recalled though these same colours and sounds, in their capacity of sense, have ceased to be seen and heard. This other form of experience is what we ordinarily call imagination. (Collingwood 1979, 204)

If a work of art is not seen or heard in the usual way but is imagined, the sensuous and the emotional combine according to a "definite structural pattern," something more than mere substructure upon which thought of truth, falsity, or illusion rests: "The activity of consciousness . . . converts impression into idea, that is crude sensation into imagination. . . . Imagination is thus the new form which feeling takes when transformed by the activity of consciousness. . . . [Note that] consciousness is not something other than thought, it is thought itself; but it is a level of thought which is not yet intellect" (Collingwood 1979, 215).

In effect, Collingwood rejected the "cult of genius," and the whole concept of aesthetic individualism (Collingwood 1979, 315–319). The processes of perception through which aesthetic sensibility is generated may be subjective—in the sense of generating an individual idea—but the ideas of imagination do not arise spontaneously in the guise of "creative genius." Instead, Collingwood believes that the emergence of ideas is bound within the community from which they arise. The epistemology of art is as much social as it is individual, and that being the case, the artist cannot forget his audience. It is easy to see why Bateson supported Collingwood's proposals.

Collingwood went even further in suggesting that the artist's "balance of polarity" emerges from the community. The role of the audience is always collaborative in art in that the artist is concerned with expressing emotions that are mutually held by both artist and by the community.

An individualistic theory of artistic authorship is absurd; on the other hand, no community knows its own heart completely, so it falls to all artists to be spokespeople of their community, and tell their audiences, "at risk of their displeasure, the secrets of their [the community's] own hearts. . . . Art is the community's medicine for the worst disease of mind, the corruption of consciousness" (Collingwood 1979, 336).

Collingwood, like Goethe before him, is one of the few major philosophical writers of his time to approach the field of aesthetics from the viewpoint of the person or participant creating the art. Many critics take issue with Collingwood's ideas, yet, the former's perspective survived until the latter part of the twentieth century. His discussion of the relation of the artist to social imagination is a significant step forward, since it privileges a participant process instead of considering aesthetics from a Kantian spectator-oriented (though "disinterested") view.

ANGELS FEAR

It took some time for Bateson to move his own writing about aesthetic appreciation away from an artist's aesthetic into ecological aesthetics. Yet when he did so, he spoke of an ecological aesthetic embracing multiple perspectives of ecological patterns in a monistic universe of information and communication. An ecological aesthetics would always be a parallax view, and would, like any vision of an overlapping third dimension (*moiré* was the term he used) yield "depth" to an understanding of "the patterns which connect." There is much more than the imaginary experience of beauty in ecological aesthetics. The creative imaginary now has to develop a sophisticated understanding of levels to imagine a first-order, second-order, and third-order comparison. In ecological patterns, first-order comparison comes from patterns of association; second-order comparison represents a dimensional contrast of relations between part and whole, comparable to level and metalevel (as in learning about context), followed by another comparison, a third-order dimension, which contrasts the pattern of part and whole in the environment of a whole. Once the domain for an ecological aesthetics is confirmed, a modification of perception about the relation between parts will modify some fundamentals of rationalism that science posits as "universals."

Conceptually, Bateson's new thinking about holism is shorthand for "that special sort of holism generated by feedback and recursiveness" (1991, 221). A new way of seeing recursion in the realm of aesthetics could, as Bateson suggests, encourage an individual to see how the natural order embodies its own mutual causality. Thus, aesthetic ideas must be brought to an interface with Bateson's coevolutionary epistemology, and give depth to a systemic understanding of the dynamics of feedback and balance in ecosystems. By enhancing feedback reciprocities between "self" and environment, the notion of an integrated organism-plus-environment comes to be seen more clearly and precisely as the unit of survival.

In short, ecological aesthetics is a metacontext for understanding pattern, and is self-reflexive, a responsive means of engaging or uncovering patterns at the second-order or even third-order level. Thus, aesthetics uncovers the underlying order of ecology, and without the metacontext of aesthetics, he argues, any new science of ecology would be bad science.

Bateson drew the connection of beauty to ecological holism in terms of a forked riddle: "What is man that he may recognize disease or disruption or ugliness?" "What is disease or disruption or ugliness that a man may know it?" (1987, 181). The riddle's two aspects present an oscillation between form and temporal process. One side of the oscillation refers to perceptual acuity in recognizing a difference between beauty and ugliness, the other, an observer's knowledge of pathology or pattern of disease. He recognized that the pattern of the percept does not flow easily into the pattern of the concept, and numerous tensions lie in the fork of the dialogue between the two. At the outset, there are issues of perception stemming from seeming contradictions in perceiving pattern flow. Next, there is the tension between appearance (ecological change) and descriptions of "reality" applied to appearances. This set of tensions arising between nonlinear patterns of events becomes a problem of epistemology. Bateson suggests working away at the fork of their contradictions. A new interface between aesthetics and ecological epistemology will likely promote a new conception of holism and will draw us toward an awareness of beauty in a larger more inclusive system. Then the interwoven regularities of the structure, a qualitative structure, may—as in all sacred realms—become the basis for awe. By contrast, wherever scientific methods cannot satisfactorily explain systemic or holistic events, they are left unexamined, fobbed off into the realm of mystery and spirituality (1987, 50–65).

The attachment of aesthetics to the science of ecology can be read in at least two ways in *Angels Fear*. The first is that of generating metaphors of unity and providing increased aesthetic sensibility to pattern and modulation of pattern—that, in turn, becomes material for dream, poetry, and more good metaphor. The other comes about through recognizing a deep connection between epistemology and aesthetics. This arises because the recursiveness of biological and psychological order is close to the root of the notion of structure or system. The injunctions of recursion are themselves fed back recursively as a constraint to relational activity in the future—to which Berleant would add that recognition of the deep connection between epistemology and aesthetics often combines to produce "aesthetics of engagement" (Berleant 1992, 2012).

An ecological aesthetics at the very least gives insight into holistic patterns pertaining to the unity of life and provides a contrast to the fragmented patterns expressed in physicalist science. Science rarely if ever deals with wholes. Developing an ability to perceive patterns of part-whole relationships is also a means through which ecology becomes a subject that investigates itself in a rigorous way, enabling discourse about connectivity, communication, and holism to spring from within the system itself. This brings biological science into the realm of artists, a universe in which the latter are both observers and participants. In such a universe, biologists must be able to play with these differences in observer/participant perspectives.

Bateson repeats themes about how ecology is an *immanent feature* of our existence, and—unlike the observer-oriented science of physical systems—requires *participant investigation*. Recursive regularities cannot be simply read out as if they are regularities derived from the code of a control program. A new way of seeing should encourage an individual to see how the natural order exhibits its own recursive tautologies. His main point is that any science predicated on "facticity" cannot in and of itself create awareness of all the implications of ecological change. Aesthetics is one means of grasping metapattern in metacontexts, uncovering a new "totality" beyond the numerical explanations of empiricism, and providing a format where the integration of form can be freely displayed and discussed (Harries-Jones 2008).

Any thumbnail summary of *Angels Fear* is difficult to present because Bateson's style does not lead to linear summation. His style is one of presentation of layers of an argument. It includes narrative accounts, based on well-known themes in human existence such as hubris, and retelling the tragedy of Oedipus and how it is related to false determinacy of Freudian psychopathology. Oedipus is a story that Freud used to establish guilt and the psychological repression of guilt, but which Bateson uses to question the efficacy of cultural habituation of belief. Another narrative tells of Coleridge's *Rime of the Ancient Mariner,* illustrating the false determinacy of conscious control and the necessity for achieving wider awareness of the relation between the conscious and the unconscious.

An ecological aesthetics would necessarily touch upon the realm of the sacred, because of the strong association of aesthetics with an idea of completion. Yet, he was too much of a Blakean to hand over to organized religion the task of creating an ecological aesthetic as other writers of his time wished to do. Bateson regarded such a leap from scientific mechanism and materialism to an ill-considered spiritualism, or to any causality that stems from high-level gods and systems invoking transcendental belief, would be a leap out of the industrial-materialist frying pan into the ethereal fire (Bateson 1987, 50–64). Understanding unity and holism in an immanent environment, he thought, should not require an embrace of otherworldly spirituality. In other words, there is no need to abandon rationality in its entirety. Rather the primary effort should be that of getting rationalists to acknowledge the "powers" of metaphor, dream, nonconscious habit, and the imaginary. This alone would be an important advance.

FACTICITY AND "JUNK SCIENCE"

Bateson was openly gloomy about the human ability to meet the threats of ecological change, and when preparing his manuscript *Angels Fear,* he wrote, "I shall argue that the imminent threat of ecological disaster is a product of epistemological error and, even more horrible than the apathy or addiction which makes it difficult, perhaps impossible, to meet this threat with appropriate action, [and] is likewise a product of epistemological error" (Bk. Mss., box 5). We can infer from the evidence in this book that Bateson is warning that epistemological error would likely shift the continuation of virtuous circles of ecological balance toward vicious circles that would not undergo their own self-correction for a very long time.

Bateson died in July 1980. Among the many eulogies was one that occurred at the dedication ceremony of the Gregory Bateson Building in Sacramento, California. The building was completed in 1979 but opened for occupancy at a later date. The governor, Jerry Brown, a prominent supporter of Bateson's ideas then and now, made provision for the construction of this building. It was to be a flagship for the Energy Efficient Office Program in California, and Brown commissioned the California state architect, Sim van der Ryn, to design it.

The Gregory Bateson Building was the first building in California for governmental use to be built around the concept of passive solar energy. At the time of the building's dedication ceremony, van der Ryn indicated what Bateson's ideas had meant to him personally:

> We found that designing a building to save energy and work with natural flows, means designing a building that is sensitive to difference and results in a building that is better for people. We found we could consider the wall of the building not as a static two dimensional architectural element, but as a living skin that is sensitive to and adapts to differences in temperature and light. We are not adapted to live or work at temperatures or lighting that are uniform and constant. We are most alive when we experience subtle cycles of difference in our surroundings. The building itself becomes "the pattern which connects" us to the change and flow of climate, season, sun and shadow, constantly tuning our awareness of the natural cycles which support all life. Maybe this is what esthetics and beauty are all about. Maybe what we find beautiful is that which connects us to an experience of differ-ence—to an experience of the patterns of wholeness which distinguish the living world from the mere works of man.

Years later, Sim van der Ryn would report that the building had reduced energy use associated with heating, cooling, and lighting compared to typical office buildings of similar scale, by more than 75 percent. He noted, "it created quite a bit of change in the building and architectural community." However, de-spite the building's evident accomplishment, it lacked sufficient follow-up so that "the sad thing is that we've wasted 30 years. We're gonna pay big time for that. Maybe we can play catch-up but it's going to take a huge shift" (Ryn 2013).

Bateson's own gloominess about the future might have anticipated a thirty-year gap in which the quantitative climate warming models available in the 1970s and 1980s went unheeded, but it is unlikely that he could have imagined the course of events that actually followed early notification of global warm-ing. Here a course of events began that fully supported his understanding of "consciousness" as being tied to social knowledge rather than individual neurophysiology. The public either doubted or denied the consensus of science about the dangers of climate change, in spite of an overwhelming global scientific consensus supporting this conclusion. Even today, some members of the public remain unconvinced that global warming is real; that it is also human induced; and that it warrants immediate action in the form of mitigation and adaptation to a new regime of higher average global temperatures.

A counternarrative to reform scientific epistemology is one matter. Denial of the consensus of empirical science is a different matter, and it suggests something more than a combination of perceptual denial of bad news, "apathy" in the public at large plus addiction to habit. For a period from the 1980s to 2012, "facts" derived from a scientific consensus of climate scientists on the International Panel on Climate Change (IPCC) faced a counter array of "facts" put together by a very well financed campaign, devised specifically to create confusion on global climate issues. Climate change denial was orchestrated. The denial lobby presented elaborate camouflage of evidence so that the public would not automatically support decisions of the global consensus of scientists. As the former vice president of the United States Al Gore, a victim of that campaign, commented, "Like an audience entertained by a magician, we allow ourselves to be deceived by those with a stake in persuading us to ignore reality" (Gore 2014).

The extent of the massive publicity campaign to reinforce denial and doubt of scientific evidence about climate change has taken some years to uncover. One of the more scurrilous aspects of the denial campaign was to create "junk science." Junk science purported to be science, but its discourse occurred with a flourish of false references to scientific appraisal (Hoggan and Littlemore 2009). People who purported to be scientists, but who had no formal training in the various fields required of professional observers of climate change, challenged the IPCC Reports simply by purporting to have studied the facts. They moved from denial to delay, and began to use court action in order to silence any critics of their highly dubious claims.

They created an "echo chamber" effect, a type of effect well known as a public relations technique. They went even further by creating phony "grass roots organizations." Their reverberating network linked their phony "grass roots" organizations to their own think tanks, to public relations firms such as Hill and Knowlton, to media blogs together with sympathetic media outlets, all of which distributed contrarian information. They used their echo chamber technique against Al Gore to cast doubt on his valid "hockey stick" reconstruction of rising Earth temperatures over the last one thousand years, which the former vice president presented in his film *An Inconvenient Truth*.

When later research showed who the spokesmen of this echo chamber enterprise were, it revealed, unsurprisingly, the fingerprints of old RANDites. This was specifically the case with two of their most prominent leaders, Frederick Singer and Frederick Seitz, both of whom had advanced their careers in the aeronautical, space technology, and satellite industry during the Cold War. These and other "merchants of doubt" carried with them the ideological economics of Kenneth Arrow and Milton Friedman. A very wealthy political group drawn to the perspective of libertarianism and laissez-fair capitalism (Oreskes and Conway 2010) gave them financial support.

Bateson once remarked that camouflage is the blanking of appropriate communication (2000, 420). Camouflage is achieved by first reducing the signal/noise ratio, he said. The second tactic of camouflage is to break up the pattern and regularities in the signal, and the third stage is by introducing similar patterns in the noise. The various climate change deniers moved their campaign along accordingly. The "merchants of doubt" moved from their initial position of denial of scientific evidence about climate change, to sowing confusion through initiating mock debates on climate change issues. Then they kept on claiming that "the scientific debate remains open" long after it had been closed—that is, by breaking up the pattern of complete consensus of international science on the issue. Finally, they moved to promote any policy that ensured delay of governmental decisions on rectifying deteriorating environmental conditions, thus supporting the illusion that though there is climate change it is not sufficient to warrant immediate government action.

Behind the spokesmen, such as Singer and Seitz, lay the finances of oil, gas, coal, and tobacco companies. Up until 2009, ExxonMobil had invested more than $20 million toward the budgets of think tanks such as the George Marshall Institute, devoted to raising questions about whether any talk of climate change was sound science. This financing of "junk science" to counter the IPCC evidence was deliberate deceit, "a story of [public] betrayal, a story of selfishness, greed, and irresponsibility on an epic scale" (Hoggan and Littlemore 2009, 15).

The mass media were seemingly unable to distinguish appropriate science from the devious spin of corporate interest groups. The media excuse was based on a cavalier understanding that presenting supposed "balance" in public news about scientific questions is good media presentation. They were also influenced by those major corporations who advertised in the media but who were part of the consortium promoting global warming denials.[1] The techniques of global climate deniers included altering language so that the worrying terminology of "global warming" became "climate change" because it was less frightening, and framing issues around arguments about the extent to which human causality was a factor in climate warming.

The period around the Copenhagen Conference on Climate Change in 2009 was the most despairing year for the IPCC, the scientific consensus, and its supporters. Ipsos MORI, in a global poll, found that while three-quarters of the people in Britain still professed to be concerned about climate change, four out of ten "sometimes think climate change might not be as bad as people say." The poll also found that six out of ten agreed that "many scientific experts still question if humans are contributing to climate change." Another 20 percent were not convinced either way. The poll concluded that people in Britain were not entirely convinced (Ipsos MORI 2008). Other polls at the time reported that scarcely 10 percent of the British public regarded global climate change as a major problem facing Britain.

By the time of the Copenhagen Conference, the "junk science" campaigners had moved to questioning the integrity of the scientists putting forward models of climate warming, using stolen emails of alleged dissenters within the scientific consensus. Al Gore recounts in his book, published just before the Copenhagen Conference, that behavioral scientists had told him that the very suggestion of negative information about the future under global climate change was triggering emotional paralysis, outright denial, or, at the very least, a refusal to undertake immediate action about the issues involved. Even though the IPCC had ensured that it had used precise, nonalarmist language in its reports, laying out the facts of global warming in this manner would not work to alter attitudes (Gore 2009). Nevertheless, the IPCC had stuck to a single vision, charting with absolute scientific certainty where climate change had occurred and whether or not human activity was implicated in global climate change. An affirmative answer was finally released in September 2013. Only a few of its members, such as James Hansen, were willing to take the wider-reaching but much more alarming precautionary view about the possibilities of climate change producing its own runaway or vicious circles leading to knock-on (feedforward) effects or "tipping points" (Hansen et al. 2013).

Meanwhile, the "junk science" controversy has overshadowed changes in biodiversity, partly from climate warming and partly from human degradation of habitat. The loss of biodiversity has already had numerous indirect impacts on ecosystems. The United Nations Millennial Assessment Report, for example, states that biodiversity loss always influences the capacity of ecosystems to adjust to changing environments. The UN Millennial Assessment is not afraid to talk about tipping points. It shows how biodiversity losses result in disproportionately large and sometimes irreversible changes in ecosystem processes, influencing potential increase in disease transmission, crop failure, impacts of pests and increased pathogens. In other words, biodiversity losses result in losses in the organization of life through an unremediated feedforward in which an interconnected series of events derive from loss in species patterns of interconnection—such is the holism of ecology (Reid et al. 2005, 77).

More recently, the International Union for the Conservation of Nature (IUCN) declared in 2010, the International Year of Biodiversity, that there was an "extinction crisis" in which "the loss of this beautiful and complex natural diversity that underpins all life on the planet is a serious threat to humankind now and in the future." The IUCN recognizes that developing any tight definition of extinction is difficult since so much of the information is based on projected estimates of total speciation, and such estimates are largely anecdotal. Estimates of population diversity and extinction rates are even more uncertain than those for species, but even if the IUCN crude estimates are on the high side, it argues, they are very alarming. Of the species it has examined, 30–50 percent may be extinct before midcentury, including one out of eight birds, one out of four mammals, one out of four conifers, one out of three amphibians, and six out of seven marine turtles.

On the other side of this gloomy perspective, recent empirical research has confirmed that aesthetics is one of the few means through which human beings interact directly with their immediate ecology. In this respect, UNESCO's creation of a network of biosphere reserves around the world attests to the reality of Bateson's idea that aesthetics could enhance ideas about the holistic aspect of ecology. The 621 biosphere reserves are community projects with a mission to develop ecological diversity, advance understanding of interactions between people and the rest of nature, and build a global capacity for the management of complex socio-ecological systems. In their aesthetic aspects, they are chosen in light of their ability to inspire awe of nature. In their physical aspects, these biosphere reserves certainly challenge the belief that logging, mining, farming, road building, and suburban sprawl can continue anywhere without any checks, regardless of the loss of biospecies. Biosphere reserves also have a learning aspect to them, for they are areas where the population within them can learn about holistic ecology and demonstrate this to others. They remain "ecological islands" fragmented in space and subject to the activities and lifestyle of the population surrounding them, and their main appeal seems to lie in the contrast between their special cases as "islands" of beauty and the outside world.

Nevertheless, genuine quandaries about the complementarity of aesthetics and ecology remain because most research on the public's response to aesthetics of nature has concentrated upon landscape studies rather than ecosystem studies. The two ideas of aesthetic perception and ecosystem sustainability, one a perceptual response and the other a cognitive response, do not necessarily match, and the two different responses have been difficult to resolve. For example, the managers of national parks in the United States and Canada are mandated to provide both aesthetics for their visitors and ecological sustainability for the park's long-term continuance. They have found the conjunction difficult. A major part of their problem is that any aesthetic notion of beauty contains a large cultural component within it. Scenic beauty is itself difficult to define, and singular features of beauty are difficult to pin down. North American ideas of the beauty of nature are bound up with the cultural notion of unchanged wilderness, and with the notion that it is the primordial quality of wilderness that yields both beauty and integrity. European ideas of natural beauty are more closely bound up with scenic areas that demonstrate careful attachment to house and surround; here, beauty invokes human sweat and it is the quality of care that gives meaning to beauty. In effect, complementarities in the relationship between aesthetics and ecology are complex, layered, contextual (Sheppard and Harshaw 2001). There is no automatic association between aesthetics and ecology.

One of the inabilities to pin down aesthetic motivation mirrors the mutual-causal panorama in ecological change itself. Raw beauty requires an interweave with other dimensions such as association with religious ritual or historical memories, or the sacredness of old age such as two thousand year old trees, in order to motivate human action. As Bateson said, "We think we see, but actually we create images all unconsciously" (1987, 94), so interaction of aesthetics and ecology requires a "halo" of effects so far as human motivation is concerned. Another constraint hindering appreciation of beauty and ecosystem is biophysical. Bateson terms this "the beam in our eye." All that we see is related to perceived difference, as conditions change. Once deprived of an ability to see movement or change, actually or figuratively, we do not directly "see" ecosystems. We "see" only at a later time that the ecology has undergone a change. The point about this "beam in our eye" is that if we cannot see differences related to ecological change—that is, if we cannot see landscape movement—we find it very difficult to generate any valuations or expectations. Maps and territory become confusing, for we cannot explain to ourselves that which we "see."

This is perhaps why the beam in the eye has played into the hands of junk science for the best part of fifteen years. From a recently revised neurophysiological viewpoint, McGilchrist reports:

> As Gregory Bateson says, all knowledge has to be knowledge of distinction, and it is of something other than the self. Equally one might say that all experience is experience of difference. Even at the sensory level we cannot experience anything unless there is a change of difference: our sensory

nerves quickly "fatigue," and we become accustomed, for example, to a smell, or to a sound. Our senses respond to the difference between values— to relative, not to absolute values. . . . The fact, that knowledge comes from distinctions, implies that we can come to the understanding of the nature of any one thing, whatever it may be, only by comparison with something else we already know, and by observing the similarities and differences. However, just as everything changes its nature, however slightly, when it changes its context, what we chose to compare a thing with determines it will stand forward and which will recede. . . . The model we choose to use to understand something determines what we find. If it is the case that our understanding is the effect of the metaphors we choose, it is also the case that it is the cause: our understanding itself guides the choice of metaphor by which we understand it . . . our relationship to the world is already revealed in the metaphors we unconsciously choose to talk about it.

HONEYBEES AND BUTTERFLIES

When Bateson drew the connection of beauty to ecological holism in terms of a forked riddle—"What is man that he may recognize disease or disruption or ugliness?" "What is disease or disruption or ugliness that a man may know it?"—he was looking for answers in terms of "looking through" the beauty of aesthetic pattern, rather than simply "looking at" it. This epistemology stems from William Blake. It is a feature of science, though, that it prefers looking at the sensible activities surrounding a problem rather than looking through in order to foster the imagination about the patterns that are evoked. And to conclude, we may take the threat of species extinction as it engulfs honeybees and butterflies.

The recent rapid increase in the annual death of honeybees holds its own mystery. Not only have their deaths in large numbers included physical manifestations of dying in or near honeybee hives, but also, and more significantly, a sudden "disappearance" from their hives. The video *The Vanishing of the Bees* reports one case of a commercial honey beekeeper who, in 2008, kept forty thousand colonies of bees—a giant enterprise. His bees suddenly disappeared. All of them, more than an estimated two thousand million bees, departed their hives in about two weeks. The same phenomenon was reported from 2006 onward in many other cases throughout the United States and Canada. Two-thirds of this "managed" honeybee population died or departed their hives between 2007 and 2010, and continue to do so. The rapid loss of honeybees reported by researchers in the US Department of Agriculture received a "mystery" label, since they could find no definite cause for colony collapse disorder

(CCD). The mystery label was then repeated in various televised documentaries such as *The Silence of the Bees* (2007) and *The Vanishing of the Bees* (2009).

Honeybees, together with birds and some other insects like butterflies, are not only aesthetically attractive species but also provide "pollination services" for more than three-quarters of the one hundred staple crop plants that feed human kind—and more than 90 percent of all flowering plants in the world. When the USDA began research on the whole question of CCD, the Department, in the usual scientific tradition, concentrated their research on a single causal factor, the use of pesticides. There is no doubt that honeybees ingest pesticide and that pesticides are embedded in the tissue of each individual bee, but initially the extensive research undertaken on genetic deformation by pesticides was unable to pinpoint one particular pesticide responsible for the loss. More recently, this type of research has shown that one group of pesticides, the neonicotinoids, a class of neuroactive insecticides chemically similar to nicotine, are the most damaging.

CCD affects "managed" honeybees, rather than those small-scale producers of organic honey. "Managed" honeybees are trucked from place to place in order to pollinate trees and vegetables, including almond trees in California, cranberries in Maine, and a variety of vegetables in Florida. The use of managed bees involves one quarter of more than 2.1 million bee colonies, which means thousands of millions of individual bees. The United States had already lost almost all its feral bees, anywhere from 80 to 95 percent by 1994.

Initially, USDA research concentrated upon the physical death of honeybees and never tackled the corresponding issue of bee behavior, such as the reason why whole populations of honeybee colonies simply disappeared by flying out of their hives. Had they done so, the research would have had to comment on the uncomfortable question of the sentience of honeybees and their seeming ability to undertake collective decision making. The bulk of research continued investigating biophysical aspects in trying to determine the reason for bee disappearance, specifically researching the presence of varroa mites in the hive and the presence of nosema, a fungus, both of which were known to cause wholesale death of honeybees.

Despite the continuing concentration of effort in biophysical and biogenetic research, behavioral clues began to emerge. For example, when bees leave their honeycombs, no other bee colonies try to take over. Typically, nearby bee populations or parasites will usually raid the honey and pollen stores of colonies that for one reason or another have been deserted by honeybees, as when the queen bee decides to search for a new place to establish a hive. Yet, in the case of CCD, other honeybees and insects left the abandoned hives untouched. Another surprise was the behavior of the queen bee, which normally resides heavily protected inside the hive, but in the case of CCD is often found outside the hive unprotected by her army of bees. This suggests that there is something wrong inside the hive and in the honeycomb itself.

In fairness, after the initial studies on the impact of pesticides on honeybee genetics, the USDA did begin to consider "stress" factors. The videos mentioned above reported these considerations. *The Silence of the Bees* showed that when honeybees are being transported from place to place by truck, from Florida to California to the Northeast and back to Florida, they are either feeding on corn syrup while en route or on specific pollens when they arrive at their industrial workplace, such as the almond trees of California, rather than a mix of pollens. In these circumstances, the bees do not collect a variety of pollens required for the proper production of nectar.

Additional information, first of all anecdotal evidence made available to the USDA, suggested that the brooding honeybee colonies were not getting enough food at a time when, normally speaking, there should have been an abundant supply of pollen and nectar. One beekeeper pointed out that even though the summer bees may seem to be foraging adequately, and trick beekeepers into believing that the summer bees are bringing adequate pollen and nectar back to the brood in the hive, beekeepers did not consider whether the pollen grains that the bees are bringing back have full nutrition value. Global warming advanced the season in which first flowering occurs, and in the process might have altered the maturation cycle of pollen in plants, so that honeybees could be bringing back immature pollen to their hives.

The true significance of change in flowering dates and other environmental clues did not appear until systematic phenological data became available. In effect, phenology is critical to an understanding of how change in the timing of ecosystem events can increase the widespread possibility of maladaptation. This is particularly the case with pollinators such as honeybees, bumblebees, and butterflies. The three species are crucial "go-betweens," and in organizational terms are crucial "crossovers" in ecosystems, affecting humans, insects, and plants (Harries-Jones 2009). They are a visible case of how timing constraints impact heterarchical flow.

Definitive research on bumblebees (rather than honeybees) appeared in 2010, when university research concluded that global warming had made a significant entry into seasonal cycles. This was based on a seventeen-year-long research project conducted by a University of Toronto scientist (Thomson 2010). Thomson's research finally established a multicausal connection between bees, first flowering, food resources, and climate warming. That same year, in 2010, research in France showed that diet did indeed have an effect on the immune system of honeybees. Honeybees rely on pollen and nectar for protein, vitamins, fats, and minerals. Typically, honeybees can always find enough mix of pollens to meet their dietary needs and to get them through a normal life cycle. The nectar that they bring back to the hive is also food for brooding bees, the next generation. The mix is also their "health food." Yet, the French study under Alaux reports, "few studies have investigated the relationship between protein nutrition and immunity in insects. . . . [Our] results suggest a link between protein nutrition and immunity in honeybees [specifically] glucose oxidase (GOX) activity, which enables bees to sterilize colony and brood food, as a parameter of social immunity" (Alaux et al. 2010, 562).

The analysis of Alaux and his colleagues provides both objective and some subjective understanding of honeybees. Objectively the bees were unable to control their natural parasite, varroa mites, due to their weakened ability to produce the type of antibiotics required for use in their hive that their foraging of pollen and nectar normally provided. Their weakened immune systems made them unable to resist the infection. Subjectively, the honeybees "knew" that they were undernourished and ill despite the fact that they were retrieving pollen and nectar from plants in their usual manner. The situation left no alternatives for the bees. They flew out to die. Flying out to die is what honeybee worker bees do when they come to the end of their working existence, but in this case their flying out occurred before the end of their honeybee life cycle and constitutes a very different phenomenon, that of a rapid evacuation of the colony.

Herein lies a cautionary tale from an insect that has always given humanity memorable metaphors of beauty, complexity, busyness, and the good life of fertility and honey. As communicative coordinators in ecosystems, their sudden decline could cause wide-scale food crop shortages long before humanity feels the full force of the destructive effects of global warming. Bateson himself had proposed that maladaptation resulting from changes in the relationship of organism to environment will be characterized by a collapse in the communicative order first of all, before physical degradation of the ecosystem occurs: "It is my personal belief that the [bio]entropic budget limits events within biological systems long before the energy budget starts to pinch" (MC, 23/01/1969, 1019–63a). It was Bateson's thesis that biological systems control the most dangerous lethal variables, such as sudden drops or increases in thermodynamic energy, through the presence of homeostatic circuits. These transform adaptation to cycling change through less lethal variables (circuits of information). Against these advantages to living systems, the major problem of ecosystems oriented toward change in less lethal variables is that their information pathways have difficulty in correcting minor pathologies.

A recent call by the David Suzuki Foundation meets the criterion of mixing aesthetics with science in an imaginative way. The Foundation has put out a call for "citizen scientists" to help "crowd source" a corridor for birds, bees, and butterflies through a large North American city. The Foundation wishes to undertake a restoration of both honeybee and monarch butterfly populations. Its program does not require intricate science, but rather straightforward action on the part of people attracted aesthetically to the pollinators. The first major activity, in both cases, is to provide appropriate food. For monarch butterflies, the first and foremost requirement is the replacement of the milkweed, the species it feeds upon, and then, perhaps, the planting of more butterfly bushes. For honeybees, this means planting a variety of plants in clumps throughout whatever gardens are available in the city, plus the building of a small "bee bath" to hold water. "Citizen scientists" can then enter the realm of science by counting butterflies, engaging in phenology,[2] monitoring larval populations, and noting any disease in the monarchs.

The cautionary tale is also about heterarchy in a world of difference, and I will give the last words to a biologist from biosemiotics, Luis Bruni, to end this volume. Bruni reports (2008, 105–106) that Bateson's concept of context-dependent information in biological systems is a final departure from the paradoxes of the physicalist account of information and has opened up an new epistemological path that sees biology as the science of sensing. There is now an integrative agenda developing in biology in which the flow of information includes the genetic and epigenetic and physiological level within organisms and the flow of information between organisms of the same or different species. As Bateson had indicated, there are two metasystems of correspondences, the genome and biodiversity. Biodiversity is the library for the ecological systems of correspondences involved in the development and organization of ecosystems. If we destroy information—that is, if we interrupt networks—we destroy the generation capacity of the ecosystem. Conversely, if we disable ecosystem-function, the information loses its sense; there will be no context for its interpretation. So besides our taxonomy of species, says Bruni, we should develop a taxonomy of information circuits. Perhaps citizen science will help in such a project. In any event, as Bateson pointed out, "The pattern is the thing."

APPENDIX

A Context Lexicon

The Shannon-Weaver version of information was to remain the standard version in academic studies of information for the next forty years or so. As Weaver said, the theory was built up from its fundamental elements. From a communications-engineering point of view, any communication system may be reduced to fundamental elements. In telephony, the signal is a varying electric current, and the channel is a wire. In speech, the signal is varying sound pressure, and the channel is the air. Frequently, things not intended by the information source are impressed on the signal. The static of radio is one example; distortion in telephony is another. All these additions may be called "noise," which is detrimental to "communication" in a channel.

> The transmitter changes a message into a signal which is sent over the communication channel to the receiver. The receiver is a sort of inverse transmitter, changing the transmitted signal back into a message, and handing this message on to the destination. When I talk to you, my brain is the information source, yours the destination; my vocal system is the transmitter, and your ear with the eighth nerve is the receiver. In the process of transmitting the signal, it is unfortunately characteristic that certain things not intended by the information source are added to the signal. These unwanted additions may be distortions of sound (in telephony, for example), or static (in radio), or distortions in the shape or shading of a picture (television), or errors in transmission (telegraphy or facsimile). All these changes in the signal may be called noise. (Weaver 1966, 17)

Weaver's model dealt with various concepts, among them information source, transmitter, noise, channel, message, receiver, channel, information destination, encode, and decode. It broke down the elements of his Sender-Receiver transmission version of information thus:

Sender: The originator of message or the information source selects desire message.

Encoder: The transmitter, which converts the message into signals. The sender's messages are converted into signals like waves or forms of binary data that are compactable to transmit the messages. For example, in telephone messaging, the voice is converted into wave signals transmitted through cables.

Decoder: The reception place of the signal converts signals into message, and is a reversal of the process of encoding. The messages are transferred from encoder to decoder through a channel. Weaver's model deals only with noises that affect the messages or signals from sources external to the channel.

Receiver: The destination of the message from sender. Based on the decoded message, the receiver will give feedback to the sender.

The Shannon-Weaver conception of information had this specific circumstance in mind: a discrete message-in-a-channel can always be identified, especially if it is a command from sender-to-receiver. Yet, their definition of information had no reference to meaning, because information was determined in relation to what the sender-receiver could say in the presence of noise, rather than what they actually did say. Messaging in the Shannon-Weaver definition is therefore not the equivalent of communication in the ordinary sense of that word, namely, engaging in a meaningful exchange; it came nearest to "meaning" in expressing what one could say in the environment of a noisy world, that is, "information is associated with the amount of freedom of choice we have in constructing messages."

When meaning is in doubt, as it usually is in any conversational communication between people, it requires knowing the "aboutness" of the communication in order to grasp its meaning. Bateson would now take upon the task of including the aboutness of the meaning in the message. This enabled him to think through a cybernetic-cum-communicational version of context. His writing produced a lexicon, covering not only differences in general characteristics between matter-energy settings and informational-communicative settings, but, in making feedback an essential attribute of communication, he created a new methodology through interactive settings. The methodology incorporated cybernetics' own nonlinear model of causality, which turned the methodology of causality inside out and rotated many other features of positivism upside down.

For example, as we will see in the table that follows, balance or invariance or conservation of energy is obtained through external but reversible pushes and pulls in matter-energy relations. In contrast, the movement of information is irreversible, but total invariance can never be obtained; rather, order is constrained through self-regulation. Self-regulated order occurs through continual change. As McNeil says, systemic informational order rests by changing, so that any informational order itself s only relatively invariant.

The "context lexicon" tabulated here is adapted from Wilden's writings (1985, 1987). While Wilden was preparing a proposal for his new book, later published as *System and Structure* (1972), he wrote to Bateson,

> It is no overstatement to say that in the cybernetic world everything is upside down. Instead of (linear) causality one deals with restraint, instead of entities, one is concerned with relationship. In the place of "individual" one finds "ecological systems," in the place of objects, "structures," in the place of facts, "messages," in the place of transfer of energy, the transmission of information . . . it is possible to integrate in a general way the emergent theory of evolution, the dialectical view of history and the behaviour of social systems . . . [but] if left to an overly technological approach lacking a critically humanistic emphasis, cybernetics and systems theory are wide open to positively dangerous ideology. (MC, 21/10/1969, 1497–7a)

To which Bateson replied in a letter to a third party: "I believe that our survival probably depends upon the sort of shift in epistemology for which Tony [Wilden] is arguing. But as he points out, this shift [to the study of information] owes its inception to the engineers and the prospect is grim if we let the engineers apply their thinking to the humanistic problems—and they are already doing this." The implication here is that the engineers always seek certainty and control of the systems they create and are prepared to act like cybernauts or, in regard to nature, are prepared to act like gods.

A Context Lexicon

A (excludes B)	(includes A as both A and B)
Matter-Energy Lexicon	Context Lexicon

General Characteristics

a. matter-energy is conserved	Information is created and destroyed
b. organization of matter-energy relations is deemed to be reversible	Organization of matter-energy relations by information is irreversible
c. random states of infinite possibilities with equilibrial order internal to system	Order through self-regulation; imaginary selection of improbable probabilities and oscillation

Causality

a. determinism expressed through numerical or statistical analysis of random probabilities	Relative indeterminism—"order from Noise"—and stochastic probability
b. straight linear causality and linear proportionality	Nonlinear cybernetics: feedback and feedforward
c. causality considered from time past to time present	Nonlinear causality considered from time present to time future (anticipation)
d. physical laws of nature	Codes of messages; rules of sensibility and perception (habits)

Descriptive Terms

a. mass, inertia, force, momentum, charge	Codes, messages, noise, information redundancy (pattern repetition)
b. mechanism must be known	mutual causal processes of self-differentiation and self-reproduction
c. atomistic; the system is the sum of its parts	Parts-whole holism; the system is other than its parts
d. only the physical level of complexity is "real"	Complexity is physical ("real") and symbolic and imaginary

Shape or Form

a. context free actually or potentially	Context constrained; context sensitive
b. few or no levels	Many levels of organization in apparent heterarchy

General Methods of Description

a. analytic, taking apart	Integral system-oriented; part-whole transforms at interface
b. isolate variables	Recognizing codes
c. either/or logic, digital yes/no; binary oppositions with quantitative measurement of analog codes	Many-valued, many-level order
d. Fragmented (bottom-up) analysis running from simple to complex	pattern of recursions are bimodal: pattern recognition is top-down and bottom-up when mapping

Dynamics

a. "force"/"power" using Newtonian causality	Constraints vis-à-vis immanent goal-seeking; second-order cybernetics
b. action and reaction	Report plus command in message produce questions in response; formation of coalitions

| c. objects and relations between entities | Patterns plus second-order patterns; virtual pattern relations between pattern relations ("in-between") |

Environment

| a. closed system not dependent on environment | Open system dependent on environment for maintaining or increasing order |
| b. closed system tending toward increasing disorder or positive entropy increase in system | recursive cycles induce neutral entropy within system by expelling entropy to environment when remaining flexible |

INFORMATION TRIGGERS CONTEXT

Wilden makes important comments about the two columns in the "context lexicon," of which the first is that the contrast and comparison evident between the two columns should not be perceived dualistically. That is to say, they should not be read as being in binary opposition, or as either/or dichotomies. He argues that the more traditional matter-energy perspective (column A) is neither rejected nor refuted by column B of context theory, and instead should be looked at as another level in a discussion of context. The vocabulary of column A should always be used in those cases whenever it works since information is always carried on some matter-energy conduit, yet is separate from it in terms of a more abstract difference at the context level. Information triggers activity in any matter-energy system, so that the context lexicon does not present a "dual" so much as levels of contrasts in which typical features (of column B) trigger matter-energy characteristics. Column B, as Wilden points out, is the basic vocabulary used to gain a perspective on information-with-meaning, in cybernetic, semiotic, and ecosystemic order (Wilden 1987, 314–317).

There is always a temptation to polarize the difference and define some type of contradiction between column A and column B. The best practice is to look at the contrast as a difference in variable intensity rather than difference in type, that is, as analog rather than digital. Thus, contrasts in column B could be a difference that emerges through (1) rotation or (2) a gestalt enlargement of the matter-energy perspectives exhibited in column. The matter-energy perspective is reductionist in column A in that it reduces qualities to quantities, and reduces complexity to simplicity. For example, column A excludes information that could be called "imaginary" although the imaginary is essential to our own visions (though such visions can be perverted, as is evident in pathology).

The "both-and" (A plus B), which are conjunctions of the second column, are intended to bring about an enlargement in the vocabulary of science. For example, matter-energy relations are described and analyzed in linear arrays. The shape or form of information systems is that they have many levels of organization, arranged in an apparent ladder of levels, and have feedback information, which is nonlinear and recursive. The shape of matter-energy systems has few or no levels and is a flatland, or "level playing field," by contrast. Information processes trigger the export of degraded energy to the environment. Matter-energy relations are reversible, while informational relations are irreversible. Another noticeable difference is that information is continually created and destroyed, while matter and energy are conserved around stability points.

When considering entropy production in physical systems, the second law of thermodynamics states that there is always an increase in positive entropy inside a closed system. When considering entropy production in organic systems, however, these systems are open to the influx both of matter-energy and also of information. As a result of the presence of both characteristics, entropy production is not the same, for a matter-energy system merely runs down, that is, degrades and becomes a homogeneous order as it does so. Organic systems, on the other hand, are open both to their biophysical environment and to information, for they are "open" systems that transport entropy to their environment as waste. This is a means whereby they can increase in variety and diversity—creating organization by taking in other species' waste and using that waste to sustain different life forms—as in the case of the dung beetle, which lives off the droppings from elephants.

From a matter-energy perspective, straight-line or linear causality proceeds from origins in past time to conditions in a present state. The variables used are quantitative—as in mass, inertia, momentum, force, or energy. Change in condition is assessed through random statistical probabilities, and here the descriptive methods used to make determinations about change are also well known: first, isolate the variables; second, take apart the component parts; and third, describe the function of the parts and how they make up the whole. It should be reasonable to assume that social systems will be far less determinate than physical systems. Social systems are material in composition, but because they contain living beings, they are highly informational as well. Yet, in the period when the Shannon-Weaver information theory was devised, the assumption was that measurement of social activity could be of the same, or similar to, measurements used in physical systems.

Take the case of "coding." Code recognition is the first step in understanding the meaning of any informational message. A "code," in the manner of a computer "code," can be physically altered, and there are always traces of its initial physical encryption. Where the information reference is primarily semiotic, that is to say, defined primarily with reference to other signs or to a pattern that reveals the meaning of a sign, then the supposition of a physical reference to the original "encryption" of a semiotic code is misleading. A semiotic code relates to some form of established rule in which there is a coupling of some items in one system, (1) to another system or systems, so that (2) a given array of signals will refer to some defined relationship or other correspondence between them. The same is true of context.

These defined relationships or correspondence do not derive from any physical bond. The bond of "blood," for example, is much more than a bond of affinity derived from biochemical similarities. Yet, as we all know, such bonds are often so emotionally strong that they are talked of as analogous to a physical state of affairs. A coding rule might also be symbolic, in which case the code is likely to be polysemic, that is, to have several alternate meanings. Or a coding rule might be purely imaginary, a one-time interchangeable sequence, as in the case of children at play. Here, rules about the coupling of relationship have no representation to any physical object whatsoever.

ACKNOWLEDGMENTS

I did not provide a summarizing conclusion at the end of my first book on Gregory Bateson, *A Recursive Vision: Ecological Understanding and Gregory Bateson*, a book based on his major archive. I felt I could not. Since that publication in 1995, such a would-be conclusion has become more evident. Bateson warns that while Western civilization has regarded its scientific and technological prowess as its major cultural virtue, the practitioners of a society's major virtues have become subject to what the Greeks called *hubris*. It is time, Bateson said, to undertake a root examination of what sort of society we are—or, as he put it, "what goes into the what?" If we do not revamp our materialist premises and turn our understanding to the broader integration of the whole and place ourselves, and our framework of knowledge—our epistemology—within an immanent ecological whole, we will not survive. This book, then, is about Bateson's counternarrative, presenting a transdisciplinary pattern integrating culture, communication, mental pathology, biology, evolution, and aesthetics, the notion of their connectedness and of the relationship between those connections, rather than the relationship between things. In Bateson's view, the way aesthetics connects with moral sense raises a moral divide between creativity and annihilation through continuing the destructive effects of biocide on lands and in oceans, seas, and atmosphere.

Many people have helped me along this path of understanding, from the many who have had email correspondence with me to those who have invited me to inspiring conferences as a result or asked me to contribute to their journals or books. In this respect, I would like specifically to thank Mary Catherine Bateson, Inge Britt-Krause, Soren Brier, Thomas Hylland Erikssen, and Don McKay. I would like to thank Don McNeil for teaching me how to grapple with the abstract notions of "heterarchy" and toroidal topology, and their necessary inclusion in any study of systems. I would like to give specific thanks to members of the Biosemiotics group, Jesper Hoffmeyer, Don Favareau, Günther Witzany, Yair Neuman, and Eliseo Fernandez. I would like to thank three former students of Gregory Bateson's, Phillip Guddemi, Frederick Ware, Jim Eicher, for expressing their in situ opinion of his teaching practices, and for Nora Bateson's recent organization of the Bateson Big Ideas Group. And thanks, too, to Brian Freer for his research on the documentation of Gregory Bateson.

My thanks go to the anonymous reviewers of a first draft who presented me with several crucial suggestions about its organization and to the reviewers at Fordham University Press, especially Jeff Bloom and Henry Sussman. To work with the enthusiasm at Fordham with Tom Lay and those concerned with the series on Meaning Systems has been a great pleasure. My thanks, too, to Angela Pietrobon for her copyediting and her indexing assistance. My friend Ian Martin and my wife, Rosalind Marguerite Gill, both had to suffer the pains of hearing about this project on too many occasions but gave ongoing advice. Were it not the fact that they are both considerable authors themselves, I am sure I would not have such forbearance.

NOTES

INTRODUCTION: A SEARCH FOR PATTERN

1. The conundrum over hybridity persisted until recently, when the function of exons in the genes became a suggested resolution. William Bateson had thought constraints to genetic performance giving rise to sterility might lie in the chromosome, but in the era before the unpacking of the gene by molecular biology (from the 1950s to the late 1990s), no biologist could find this chromosomal "residue" constraining the gene (Cock and Forsdyke 2008, 672).

2. A recent fictional account of their courtship in New Guinea and subsequently, based on archival material, can be found in Lily King's novel *Euphoria* (2014).

3. A new documentary film entitled *Ecology of Mind*, produced and directed by Nora Bateson, the younger daughter of Gregory Bateson, explores her father's themes through a humorous, loving, and deeply personal framework. Connected segments of Bateson's life, bring forward his important ideas such as pattern, relationship, cybernetics, "difference that makes a difference," and double bind. It is a visual walk through of his major terms and ideas, rather like a walk through an art gallery, together with people from various disciplines who talk about Bateson's relevance to their own work. The DVD proposes that multiple levels of "mind" are a good metaphor for perception of ecology. When "mind" is perceived only as networks of neurons instead of formal patterns of relations, the relationship between humanity and nature appears to be highly separated.

1. CULTURE: A FIRST LOOK AT DIFFERENCE

1. Anthropology in general has neglected the influence of Bartlett on Bateson. An exception is Maurice Bloch, who traces Bartlett's ideas to the latter's intellectual predecessor at Cambridge, W. H. R. Rivers, but argues that many anthropologists have ignored the essential contribution of a study of memory to the study of culture (Bloch 1996, 108, 362). In recent years, Rivers's rebuttal of behaviorism has become a subject of film, which in turn is based on the remarkable literary trilogy on Rivers by Pat Barker.

2. "Connecting things or events together which occur at different points in a succession is obviously and directly a process of filling up an interval in the evidence. And here also the interval is filled in accordance with standard relationships that are somehow immediately appreciated. . . . Suppose, however, the gap is very large. The situation has changed. There is some stage in an increasing size of interval at which it becomes reasonable or necessary to fill it up by a number of steps each related to the next. . . . Here we appear bound to work within a particular system, and every step follows necessarily upon every other and even the order of steps is nearly, though not quite, necessary. It would seem, then, that thinking can be treated as a process of filling up gaps by a series of steps, in accordance with incomplete evidence that is appreciated by our senses; that the steps must be consistent with one another; that when we have to use the materials of day by day experience there is a good deal of freedom about the step we have to pick out to start from, and for this we need imagination" (Bartlett 1951, 123 ff.).

3. Their eventual choice of a highland village, where they stayed for two years, Bajoeng Gede (now Bayunggede), for Mead, "was one of those lucky accidents that have accompanied me all my life," in that instead of the clay walls which cut off other houses and households from sight, this village consisted of bamboo fencing so that she could catch a glimpse of what was going on in a courtyard without having to actually enter a house. Entering a house always required "courtesies and gifts of refreshment on every occasion when one entered a courtyard." But the mountain people "were a dour people and suspicious of strangers; they lacked the easy openness to any patron that characterized the people of the lowlands" (Mead 1972, 254).

4. Bateson had noted that there had been wars in precolonial times and that people were killed in these wars, but had stressed that wars were rare events. Bateson might have discussed how and why the Dutch colonial administration had created a setting in which violence was rare. When the Dutch came to occupy Bali, the island experienced an extraordinary event in its history. The ruling court and its courtiers responded to threats of violence by the army of the Dutch colonials who had come to take over the island in the early 1900s by parading down to the shore and committing mass suicide. Balinese dignitaries walked unarmed, offering themselves in front of the guns of the Dutch soldiers and then began turning their own *kris* daggers upon themselves. So disturbed were colonial rulers to this reaction that they decided to keep Balinese traditional customs alive. On the neighboring island of Java and elsewhere, they transformed the administration of the indigenous state through active intervention (Pringle 2004). The postcolonial era initiated serious factionalism and bloodletting. Following the fall of the left-wing Sukarno regime in Indonesia (of which Bali had become part) in 1967, a conflict between the left-wing heroes of Bali's independence and the right wing that came into power under Suharto resulted in a massacre by Suharto of left-wing Balinese. The situation pitted Bali resident against Bali resident and gave rise to bloodshed that was as horrific in population per capita killings as the genocide that occurred later under the Pol Pot regime in Cambodia during the 1970s. Since there was no 'truth and reconciliation' campaign following the demise of Suharto, today,

families of Communist Party members who were the targets of the massacre are still not reconciled with the families of the perpetrators.

5. Hildred Geertz was to question the "gods as children" comment to which Bateson made reference. She wondered whether Bateson was using the Indonesian translation of *anak sakti*, which would indeed translate into a "child" of spiritual power. Geertz discusses the upside-down god configured on the cover of this book, which Bateson and Mead commissioned from the Bali artist Ida Magus Made Togog (Geertz 1994, 21).

6. Her own use of Freudian psychodynamics is particularly prominent when she put together another book entitled *Growth and Culture: A Photographic Study of Balinese Childhood*, which is dependent on Arnold Gesell's revision of Freud's developmental stages of childhood. Bateson did not contribute to this volume (Mead and Macgregor 1951).

2. A SCIENCE OF DECENCY

1. These included a variety of anthropologists and cultural writers, including Franz Boas, Ruth Benedict, Edward Sapir, Erich Fromm, John Dollard, and Geoffrey Gorer. Mead lists these as those who had the greatest influence on her own thinking at the time in the appendix for one of the few books she never completed, to be entitled *Learning to Live in One World*.

2. The initial refusal of the United States to become engaged in a war in Europe turned Bateson toward criticizing British efforts on getting out a message about the immanent dangers, stating that the embassy itself had "a miniature information service with no money, a few good people and no backing from home." When writing to his relatives in England, he pleaded that even "a very few hundred English people in this country picked for their liking of America and primed with some knowledge of the difference between England and America plus about the cost of a battleship would have been a good investment" (Letters, LOC, 15/03/1942).

3. Sir John Cockroft took charge of the Canadian Atomic Energy project in 1944 and became director of the Montreal and Chalk River Laboratories. He won the Nobel Prize in 1951 for his preliminary work on splitting the atom.

4. Osborn was a most unlikely choice to run such a project, since he was best noted in sociological circles for his support of eugenics. As secretary of the American Eugenics Society, Osborn followed developments in eugenics in Germany with great interest. He had written a report in 1937 summarizing developments in the German sterilization program, and applauding its successes, had noted how such progress, in the form of rapid social change, could only be made under a dictator. Eugenics fell into disgrace in the immediate postwar period, after the revelations of Auschwitz and Belsen. In the eyes of many of his contemporaries, Osborn had simply shifted his focus slightly from a biological eugenics toward a social eugenics. His support for the use of statistics in social science to direct and control human behavior could be seen in conjunction with his earlier support of the use of statistics in the German sterilization program Needless to say, "social eugenics" did not represent the outcome that Bateson wished for as a

result of "professional" social scientists entering the field of National Morale. Osborn would write the foreword to *The American Soldier* (Stouffer et al. 1949–50, ix).

5. During the years before he was accepted into active service, Bateson spent some time on film projects, including work on his own film ethnography of trance and dance in Bali, and film reviews of Nazi propaganda movies, such as *Hitlerjunger Quex*.

6. The British delegated some responsibilities to colonial officers in Burma and elsewhere through the means of customary law, but not on any equal basis. Under "indirect rule," the indigenous leaders, appointed by the British, carried out certain functions of colonial law and order in conjunction with customary law. Indigenous leaders, effectively cultural headmen, would collect taxes, appoint subordinates, aid the police in the administration of law and order, hear court cases with regard to theft or marriage agreements, and prosecute witchcraft cases. However, no customary ruler was ever the equal of a colonial administration head.

7. This account comes from Price, who obtained it from CIA files under a Freedom of Information request (Price 2008, 3–6).

8. The well-known English politician David Owen has enumerated out of his own political experience the characteristics of hubris as being messianic manner; excessive confidence in own judgment; contempt for advice; exaggerated self-belief, bordering on omnipotence; and a belief that one is accountable solely to history or god, all of which leads to loss of contact with reality often associated with progressive isolation, restlessness, recklessness, and impulsiveness.

9. *Readings* offers a primer in the use of quantitative data with examples from Army surveys. Yet none of the empirical studies presented, nor the editorial discussion of the book, went beyond immediate results to discuss whether or not qualitative evidence and quantitative evidence had been drawn from two very different levels of abstraction, and so distorted the way in which dominance and subordination in society was perceived. For example, studies of class formation and racial antagonism showed through qualitative evidence that crossovers of opinion between black and white on race issues were next to impossible in those days. Race prejudice created a qualitative distinction, a visible gap between the majority of black people and whites. Articles in *Readings* showed in attitude surveys on class distinction that middle class could identify with the working class on some issues, and vice versa.

10. His posting late in life to the University of California at Santa Cruz Department of Anthropology during the 1970s was a part-time shared appointment with the History of Consciousness Studies.

4. WHY WE SEE IN OUTLINES

1. Bateson acknowledges that this sort of explanation can verge on runaway, as "the possibility of differences between differences and differences which are differently effective and differently meaningful according to the network in which they exist is a path towards an epistemology of gestalt psychology, and this clumping of news of a difference becomes specially true of the mind when it . . . evolves language."

INTERLUDE: FROM CULTURAL STRUCTURES TO STRUCTURE IN ECOLOGY

1. Neuman notes that when an organism gradually ceases to respond automatically to the mood-signs of another and comes to recognize the sign as a signal that can be trusted, distrusted, falsified, denied, amplified, or corrected, this metalinguistic ability, as Bateson pointed out, is an important step in the development of communication. Thus the signal of a relationship lies in the sign, metalanguage necessarily accompanies language, and the metalinguistic level always qualifies the referential power of the sign. "Through the meta-language the sign is established as a unique entity that exists in between the two realms . . . through the meta-linguistic level the sign is in a state of superposition, that endows it with the power to transform the closed semiotic system into a non-semiotic (relational?) realm without it being an integral part of any system" (Neuman 2008, 67–68).

2. Bateson discusses these ideas in relation to his father's work on the appearance of limbs, or nonappearance of limbs in organic monstrosities. He also considers how symmetries and asymmetry of shape can occur in plants through different morphology of the fork of the plant, which enables a flower to be not radially but bilaterally symmetrical (2000, 395).

6. PATTERN AND PROCESS

1. For an interesting account of the unraveling of Robert May's mistaken propositions about complexity, see the discussion in Buchanan 2002 about the theory and history of networks, both social networks and organic networks.

2. The switch in thinking occurred after Bateson's death. The notion of information theory current today is that if an ecosystem is *not* treated in terms of mechanical dynamics nor as deterministic systems, then information theory can be retained to measure connectivity in a network of pathways in an ecosystem. In other words, materialist-minded ecologists now find it possible to proceed with measurements of information in ecosystem networks without assuming that their measurements reflect a deterministic situation with respect to diversity and stability. This change in thinking is a small shift compared to Bateson's proposals about bioentropy.

7. A POSTGENOMIC VIEW

1. A postgenomic view expands the protein's role into an element in a network of protein-protein interactions as well, in which it has a contextual or cellular functions within functional macromolecules, with many examples of patterned beauty (Goodwin 2008, 147–148).

2. McNeil's advanced work is not yet published and exists on a CD entitled 'Goings On With Systems.' A preliminary publication, 'What's Going On with the Topology of Recursion,' is available online at: www.library.utoronto.ca/see/SEED/Vol4-1/McNeil .htm. It is a useful accompaniment to the following discussion.

3. Quoted in Ames and Hall 2003, 91.

8. TOWARD THE SEMIOSPHERE

1. This level of purposiveness Millikan terms "affordances," a term borrowed from the environmental psychologist J. J. Gibson.

2. Note that an *interpretant* should be distinguished from an *interpreter*, for in any Peircian discussion they are not necessarily one and the same. An interpretant such as "habit" belongs to the propositional order of a "sign." In semiogenic living systems, participants examine regularities in semiotic behavior and interaction. Regularities are derived from habits that provide *interpretants*, by which a sign or sign-vehicle is connected to its object.

9. ECOLOGICAL AESTHETICS AS METAPATTERN

1. Such camouflage has continued to be highly successful. A poll released about the melting Arctic sea ice in Canada in 2012 suggested that a significant number of Canadians still have trust issues with certain areas of climate change science, despite regular visual reports on disappearing ice in Canadian Arctic regions. A total of 28 percent of Canadians said that they distrusted or somewhat distrusted research on that issue. The Canadian government was also dismissive of the dangers of climate change. The polls in the United States also reported similar numbers of climate change doubters (*Harpers* 2013).

2. Phenology is the study of the time of first flowering in spring, a study that used to be widespread in the nineteenth century among natural historians using relative simple techniques of gathering data. Phenology has become important once again because climate change is altering flowering times worldwide. In most cases, plants are flowering earlier, although patterns of early flowering can be species specific, or depend on local conditions. Understanding of the timing of ecological events would lead to a better understanding of ecosystems.

SELECT BIBLIOGRAPHY

1. PUBLISHED WORK BY GREGORY BATESON AND MARY CATHERINE BATESON

Note: The numbering used in this select bibliography accords with that of the complete bibliography of Bateson's work published in *A Sacred Unity* (Bateson 1991, 314–336).

Bateson, Gregory. 1941. "Experiments in Thinking About Observed Ethnological Material." *Philosophy of Science* 8 (1).

———. 1943a. "Discussion: The Science of Decency." *Philosophy of Science* 10 (2): 140–142.

———. 1943b. "An Analysis of the Film Hitlerjunge Quex (1933)." In *The Study of Culture at a Distance*, edited by Margaret Mead and Rhoda Metraux, 302–314. Chicago: Chicago University Press.

———. 1946a. "The Pattern of an Armaments Race: An Anthropological Approach—Part I." *Bulletin of the Atomic Scientists* 2 (5–6): 25–28.

———. 1946b. "The Pattern of an Armaments Race: An Anthropological Approach—Part II." *Bulletin of the Atomic Scientists* 2 (5–6): 25–28.

———. 1946c. "Art of the South Seas." *Art Bulletin* 28 (2): 119–123.

———. 1958. *Naven: A Survey of the Problems Suggested by a Composite Picture of the Culture of a New Guinea Tribe Drawn from Three Points of View.* Cambridge: Cambridge University Press. Originally published 1936.

———. 1961. "Introduction." *Perceval's Narrative: A Patient's Account of His Psychosis, 1830–32.* Stanford: Stanford University Press.

———. 1968a. "Redundancy and Coding." In *Animal Communication: Techniques of Study and Results of Research*, edited by Thomas A. Sebeok. Bloomington, Indiana: Indiana University Press.

———. 1979a. *Mind and Nature: A Necessary Unity.* New York: Dutton.

———. 1979b. "Nuclear Armament as Epistemological Error: Letters to the Board of Regents." *Zero* 3: 34–41.

———. 1980. "The Oak Beams of New College, Oxford." In *The Next Whole Earth Catalogue*, edited by Stewart Brand, 77. New York: Random House.

———. 1991. *A Sacred Unity: Further Steps to an Ecology of Mind,* edited by Rodney L. Donaldson. New York: HarperCollins.

———. 2000. *Steps to an Ecology of Mind*. Chicago, University of Chicago Press. 2000. Originally published 1972.

Bateson, Gregory, and Mary Catherine Bateson. 1987. *Angels Fear: Towards an Epistemology of the Sacred*. New York and London: Macmillan.

Bateson, Gregory, and Claire Holt. 1944. "Form and Function of the Dance in Bali." In *The Function of Dance in Human Society: A Seminar Directed by Franziska Boas*, 46–52. New York: the Boas School.

Bateson, Gregory, and Margaret Mead. 1942. *Balinese Character: A Photographic Analysis*. New York: New York Academy of Sciences.

Bateson, Gregory, and Jurgen Ruesch. 1951. *Communication: The Social Matrix of Psychiatry*. New York: Norton.

Bateson, Mary Catherine. 1972. *Our Own Metaphor: A Personal Account of a Conference on the Effects of Conscious Purpose on Human Adaptation*. New York: Knopf.

———. 1984. *With a Daughter's Eye: A Memoir of Margaret Mead and Gregory Bateson*. New York: Morrow.

2. OTHER PUBLISHED WORKS CITED

Abella, Alex. 2000. *Soldiers of Reason: The RAND Corporation and the Rise of the American Empire*. Orlando, Fla: Harcourt.com.

Alaux, Cédric, François Ducloz, Didier Crauser, and Yves Le Conte. 2010. "Diet Effects on Honeybee Immunocompetence." *Biology Letters* 6: 562–565.

Allott, Robin. 2005. "How Children Acquire Language: The Motor Theory Account." In *The Child and the World: How the Child Acquires Language, How Language Mirrors the World*, 1–48. Accessed at: www.percepp.com/langacqu.htm.

Ames, Roger T., and David L. Hall. 2003. *Dao De Jing: Making This Life Significant*. New York: Random House.

Arrow, Kenneth. 1951. *Social Choice and Individual Values*. New York: Wiley.

Baluska, Frantisek, Stefano Mancuso, and Dieter Volkmann, eds.. 2009. *Signaling in Plants*. Dordrecht: Springer.

Barnes, Julian. 1984. *Flaubert's Parrot*. London: Jonathan Cape.

Bartlett, Frederic C. 1928. *Psychology and Primitive Culture*. Cambridge: Cambridge University Press.

———. 1951. *The Mind at Work and Play*. London: George Allen Unwin.

———. 1961. *Remembering: A Study in Experimental and Social Psychology*. Cambridge: Cambridge University Press.

Beck, Ulrich. 1992. *Risk Society: Towards a New Modernity*. London: Sage Publications.

Bell, Simon. 2001. "Can a Fresh Look at the Psychology of Perception and the Philosophy of Aesthetics Help Contribute to the Better Management of Forested Landscapes?" In *Forests and Landscapes*, edited by Stephen Richard John Sheppard and Howard W. Harshaw, 125–148. New York: CABI Publishing.

Belo, Jane. 1970. *Traditional Balinese Culture: Essays*. New York: Columbia University Press.

Benedict, Ruth. 1934. *Patterns of Culture*. Boston, Mass.: Houghton Mifflin.

Bennett, John W. 1998. *Classic Anthropology: Critical Essays, 1944–1996*. New Brunswick, N.J.: Transaction Publishers.

Berger, John. 2005. "A Jerome of Photography." *Harpers Magazine* 311 (1867), 87.

Berleant, Arnold. 1992. *The Aesthetics of Environment*. Philadelphia: Temple University Press.

———. 2012. "Environmental Aesthetics." In *Art and Environment: Proceedings of Sanart III Symposium, Ankara*. Accessed November 29, 2012. www.sanart.org.tr/artenvironment/Berleant_fullpaper.pdf.

Bloch, Maurice. 1996. "Cognition" and "Memory." In *Encyclopedia of Social and Cultural Anthropology*, edited by Alan Barnard and Jonathan Spencer, 108–110, 361–363. New York: Routledge.

Bolter, Jay David. 1984. *Turing's Man: Western Culture in the Computer Age*. Chapel Hill: University of North Carolina Press.

Boulding, Kenneth. 1971. "Toward a General Theory of Growth." In Hinton and Reitz 1971, 516–524.

Bowden, Margaret A. 1990. *The Creative Mind: Myths and Mechanisms*. London: Basic Books.

Bowers, C. A. 1997. *The Culture of Denial: Why the Environmental Movement Needs a Strategy for Reforming Universities and Public Schools*. Albany: SUNY Press.

Brand, Stewart. 1974. *Two Cybernetic Frontiers*. New York: Random House.

Bredo, Eric. 1989. "Bateson's Hierarchical Theory of Learning and Communication." *Educational Theory* 39(1): 27–37.

Britt-Krause, Inge. 2003. "Learning How to Ask in Ethnography and Psychotherapy." *Anthropology and Medicine* 10:3–22.

Brockman, John. 1977. *About Bateson: Essays on Gregory Bateson*. New York: Dutton.

Bruni, Luis. 2008. "Gregory Bateson's Relevance to Current Molecular Biology." In Hoffmeyer 2008b, 93–119.

Buchanan, Mark. 2002. *Small Worlds and the Groundbreaking Theory of Networks*. New York: Norton.

Bürglin, Thomas R. 2006. "Genome Analysis and Developmental Biology," In Neumann-Held and Rehmann-Sutter 2006, 15–37.

Campbell, Jeremy. 1982. *Grammatical Man: Information, Entropy, Language, and Life*. New York: Simon & Schuster.

Capra, Fritjof. 1996. *The Web of Life: A New Scientific Understanding*. New York: Doubleday.

Carson, Rachel. 1962. *Silent Spring*. New York: Houghton Mifflin.

Carson, Robert C. 1969. *Interaction Concepts of Personality*. Chicago: Aldine.

Claverie, Jean-Michel, and Chantal Abergel. 2012. "The concept of Virus in the Post-Megavirus Era." In Witzany 2012, 187–202.

Cock, Alan, and Donald Forsdyke. 2008. *Treasure Your Exceptions: The Science and Life of William Bateson*. Dordrecht: Springer.

Collingwood, Robin George. 1979. *The Principles Of Art*. London: Oxford University Press.

Contributions of Working Group II to the Third Assessment Report of the Intergovernmental Panel on Climate Change. Cambridge, Cambridge University Press. Working Group 2: Feedback, Interactions and Resilience. Accessed January 11, 2005. www.grida.no/climate/ipcc_tar/wg1/index.htm.

Covarrubias, Miguel. 1989. *Island of Bali*. Oxford: Oxford University Press.

Crick, Francis, and Kristof Koch. 1998. "Consciousness and Neuroscience." *Cerebral Cortex* 8: 97–107.

Currie, Mark. 2004. *Difference*. London and New York: Routledge.

Dear, Ian. 1996. *Sabotage and Subversion: The SOE and the OSS at War*. London: Cassell.

Dell, Paul F. 1982. "Beyond Homeostasis: Towards a Conception of Coherence." *Family Process* 21 (1).

———. 1985. "Understanding Bateson and Maturana: Towards a Biological Foundation for the Social Sciences." *Journal of Marital and Family Therapy* 1:1–19.

Delmer, Sefton. 1962. *Black Boomerang*. London: Secker and Warburg.

Donaldson, Rodney. 1987. *Gregory Bateson Archive: A Guide/Catalog*. 4 vols. Ann Arbor, Mich.: University Microfilms International Dissertation Information Service.

Eckermann, Johann P. 1998. *Conversations of Goethe*. Translated by John Oxenford, edited by J. K. Moorhead. London: Da Capo Press.

Eldredge, Niles, and Stephen J. Gould. 1972. "Punctuated Equilibria: An Alternative to Phyletic Graduation." In *Models in Paleobiology*, edited by T. J. M. Schopf, 82–115. San Francisco: Freeman Cooper.

Ellis, Donald. G. 1981. "The Epistemology of Form." In *Rigor and Imagination: Essays from the Legacy of Gregory Bateson*, edited by Carol Wilder and John H. Weakland, 215–230. New York: Praeger.

Emmeche, Claus, and J. Hoffmeyer. 1991. "From Language to Nature: The Semiotic Metaphor in Biology." *Semiotica* 84(1/2): 1–42.

Evans-Pughe, Christine. 2014. "Turing's Morphogenesis Theory Drives Research into Self-Configuring Systems." *Engineering and Technology Magazine* 9(11), November 10, 2014. Accessed December 24, 2014. http://eandt.theiet.org/magazine/2014/11/natures-blueprints-unmasked.cfm.

Favareu, Donald, ed. 2010. *Essential Readings in Biosemiotics: Anthology and Commentary*. Dordrecht: Springer.com.

Fernandez, Eliseo. 2010. "Living Is Surviving: Causation, Reproduction and Semiosis." Tenth Annual International Gatherings in Biosemiotics. Braga, Portugal, June 22–27.

Festinger, Leon. 1971. "Informal Social Communication." In Hinton and Reitz 1971, 223–232.

Flaskas, Carmel. 2002. *Family Therapy Beyond Postmodernism: Practice Challenges Theory*. New York: Brunner-Routledge.

Fortes, Meyer, and Edward Evan Evans-Pritchard, eds. 1940. *African Political Systems.* Oxford: Oxford University Press.

Foucault, Michel. 1988. *Madness and Civilization: A History of Insanity in the Age of Reason.* Translated by Richard Howard. New York: Vintage Books.

Fox Keller, Evelyn. 2006. "Beyond the Gene but Beneath the Skin." In Neumann-Held and Rehmann-Sutter 2006, 290–312.

French, John R. P., and Bertram Raven. 1959. "The Bases of Social Power." In *Studies in Social Power,* edited by Dorwin Cartwright, 150–167. Ann Arbor: University of Michigan, Institute for Social Research.

Geertz, Clifford. 1966. *Person, Time, and Conduct in Bali: An Essay in Cultural Analysis.* New Haven: Southeast Asia Studies, Yale University.

Geertz, Hildred. 1994. *Images of Power: Balinese Paintings Made for Gregory Bateson and Margaret Mead.* Honolulu: University of Hawaii Press.

Giorgi, Franco. 2012. "Agency" In *A More Developed Sign: Interpreting the Work of Jesper Hoffmeyer,* edited by Donald Favareau and Paul Cobley and Kalevi Kull, 13–16. Tartu: Tartu University Press.

Goethe, Johann W. von. 1995. *The Scientific Studies.* Translated and edited by D. Miller. Princeton: Princeton University Press.

Goffman, Erving. 1974. *Frame Analysis.* New York: Harper & Row.

Goldhammer, Eberhard von, and Joachim Paul. 2007. "The Logical Categories of Learning and Communication Reconsidered from a Polycontextural Point of View." *Kybernetes* 36(7/8): 100–1011.

Goodwin, Brian. 1994. *How the Leopard Changed Its Spots: The Evolution of Complexity.* New York: Simon & Schuster.

———. 2008. "Bateson: Biology with Meaning." In Hoffmeyer 2008b, 147–148.

Gore, Al. 2009. *Our Choice: A Plan to Solve the Climate Crisis.* Emmaus, Pa.: Rodale Books.

———. 2014. Review of Elizabeth Kolbert, *The Sixth Extinction. New York Times,* February 10.

Gould, Stephen Jay. 1989. *Wonderful Life: The Burgess Shale and the Nature of History.* New York: Norton.

Gregory, Richard L., ed. 1987. *The Oxford Companion to Mind.* Oxford: Oxford University Press.

Guattari, Félix. 2000. *The Three Ecologies.* Translated by Ian Pindar and Paul Sutton. London: Athlone Press.

Harpers Magazine. 2013. "Index," January.

Harries-Jones, Peter. 1985. "The Nuclear Winter Hypothesis: A Broadened Definition." In *Nuclear Winter and Associated Effects: A Canadian Appraisal of the Environmental Impact of Nuclear War.* Royal Society of Canada, Ottawa, January 31, 1985: 374–382.

———. 1995. *A Recursive Vision: Ecological Understanding and Gregory Bateson.* Toronto: University of Toronto Press.

———. 2008. "Gregory Bateson's 'Uncovery' of Ecological Aesthetics." In Hoffmeyer 2008b, 153–167.

———. 2009. "Honeybees, Communicative Order, and the Collapse of Ecosystems."
Journal of Biosemiotics 2:193–204.

———. 2010. "Bioentropy, Aesthetics and Meta-dualism: The Transdisclipinary Ecology of Gregory Bateson." *Entropy* 12(12): 2359–2385.

Heims, S. J. 1982. *John von Neumann and Norbert Wiener: From Mathematics to the Technologies of Life and Death.* Cambridge, Mass.: MIT Press.

Held, B. S., and E. Pols. 1985 "Rejoinder: On Contradiction." *Family Process* 24:521–524.

Hinton, Bernard, and H. Joseph Reitz, eds. 1971. *Groups and Organizations: Integrated
Readings in the Analysis of Social Behavior.* Belmont, Calif.: Wadsworth Publishing.

Hochstein, Shaul, and Morav Ahissar. 2002. "View from the Top: Hierarchies and Reverse Hierarchies in the Visual System." *Neuron* 36 (December): 791–804.

Hoffman, Lynn. 1985. "Beyond Power and Control: Towards a Second Order Systems
Therapy." In *Is Earth a Living Organism? Proceedings of a Conference at the University of
Massachusetts.* National Audubon Society Expedition Institute Paper 33.

Hoffmeyer, Jesper. 1996a. *Signs of Meaning in the Universe.* Translated by Barbara Have-
land. Bloomington: Indiana University Press.

———. 1996b. "Evolutionary Intentionality" In *The Third European Conference on Systems Science,* edited by E. Pessa, A. Montesanto, and M. P. Penna, 699–703. Rome:
Edizioni Kappa.

———. 2002. "Code Duality Revisited." *Semiotics, Evolution, Energy and Development
Journal* 2(1): 98–117.

———. 2008a. *Biosemiotics: An Examination into the Signs of Life and the Life of Signs.* Chicago: University of Scranton Press.

———, ed. 2008b. *A Legacy for Living Systems: Gregory Bateson as a Precursor to Biosemiotics.* Dordrecht: Springer.

Hofstadter, Douglas. 2007. *I Am a Strange Loop.* New York: Basic Books.

Hoggan, James, and Richard Littlemore. 2009. *Climate Change Cover-up: The Crusade to
Deny Global Warming.* Vancouver: Greystone Books.

Ipsos MORI. 2008. "Public Attitudes to Climate Change: concerned but still unconvinced." Poll, May 23–May 28 2008. Accessed May 2012. https://www.ipsos-mori.com/
researchpublications/researcharchive/2305/Public-attitudes-to-climate-change-2008
-concerned-but-still-unconvinced.aspx.

Jung, Carl G. 1965. "Septem Sermones ad Mortuos." In *Memories, Dreams and Reflections,*
edited by Amelia Jaffé, 378–390. New York: Random House.

Keeney, Bradford P. 1982a. "Ecosystemic Epistemology: Critical Implications for the
Aesthetics and Pragmatics of Family Therapy." *Family Process* 21(1): 1–20.

———. 1982b. "What Is an Epistemology of Family Therapy?" *Family Process* 21(1):
153–168.

Kemp, Gary. 2012. 'Collingwood's Aesthetics.' *Stanford Encyclopedia of Philosophy.* Accessed March 23, 2013. http://plato.stanford.edu/entries/collingwood-aesthetics.

Keynes, Geoffrey, ed. 1972. *Blake: Complete Writings.* Oxford: Oxford University Press.

King, Lily. 2014. *Euphoria: A Novel.* London: Atlantic Monthly Press.

Laing, Ronald David. 1961. *Self and Others.* Harmondsworth: Penguin Books.

———. 1967. *Politics Of Experience*. Harmondsworth: Penguin Books.

Lipset, David. 1980. *Gregory Bateson: the Legacy of a Scientist*. Englewood Cliffs, N.J.: Prentice-Hall.

Maturana, Humberto. 1975. "The Organization of the Living: A Theory of the Living Organization." *International Journal of Man-Machine Studies* 7(3): 313–332.

Maturana, Humberto R. and Francisco J Varela. 1992. *The Tree of Knowledge: The Biological Roots of Human Understanding*. Translated by Robert Paolucci. Boston: Shambhala.

May, Robert. 1973. *Stability and Complexity in Model Ecosystems*. Princeton: Princeton University Press.

McCarthy, James J., Osvaldo F. Canziani, Neil A. Leary, David D. Dokken, and Kasey S. White, eds. 2001. Climate Change 2001: Impacts, Adaptation and Vulnerability. Contribution of Working Group II to the Third Assessment Report of the Intergovernmental Panel on Climate Change (IPCC). Cambridge: Cambridge University Press. Accessed January 11, 2005. http://www.grida.no/publications/other/ipcc_tar/?src=/climate/ipcc_tar/wg2/index.htm.

McCulloch, Warren. 1965. *Embodiments of Mind*. Cambridge, Mass.: MIT Press.

McGilchrist, Iain. 2009. *The Master and His Emissary: The Divided Brain and the Making of the Western World*. New Haven and London: Yale University Press.

McNeil, Don. c. 2003. "What's Going On with the Topology of Recursion." Accessed May 2014. www.library.utoronto.ca/see/SEED/Vol4-1/McNeil.htm

Mead, Margaret. 1940. "Educative Effects of Social Environment Disclosed by Studies of Primitive Societies." In *Environment and Education*, edited by Ernest W. Burgess, W. Lloyd Warner, Franz Alexander, and Margaret Mead. Human Development Series, 1(54): 48–61. Chicago: University of Chicago Supplementary Educational Monographs.

———. 1946. *And Keep Your Powder Dry: An Anthropologist Looks At America*. New York: Morrow.

———. 1972. *Blackberry Winter: My Earlier Years*. New York: William Morrow.

———. 1977. *Letters from the Field, 1925–1975*. New York: Harper & Row.

Mead, Margaret, and Frances C. Macgregor. 1951. *Growth and Culture: A Photographic Study of Balinese Childhood*. New York: Putnam.

Merrell, Floyd. 2003. *Sensing Corporeally: Toward a Posthuman Understanding*. Toronto: University of Toronto Press.

Millikan, Ruth Garrett. 2004. *Varieties of Meaning*. Cambridge, Mass.: MIT Press.

Moreno, Jacob L. 1934. *Who Shall Survive? Foundations of Sociometry, Group Psychotherapy, and Sociodrama*. New York: Beacon House.

Neuman, Yair. 2008. *Reviving the Living: Meaning Making in Living Systems*. Amsterdam: Elsevier.

Neumann-Held, Eva, and Christof Rehmann-Sutter, eds. 2006. *Genes in Development: Re-Reading the Molecular Paradigm*. Durham, N.C.: Duke University Press.

Newcomb, Theodor, and Eugene Hartley, eds. 1947. *Readings in Social Psychology. Prepared for the Committee on the Teaching of Social Psychology*. New York: Henry Holt.

Newman, Stuart, and Gerd B. Müller. 2006. "Genes and Form: Inherency in the Evolution of Developmental Mechanisms." In Neumann-Held and Rehmann-Sutter 2006, 38–76.

Noble, David F. 1984. *Forces of Production: A Social History of Industrial Automation.* New York: Oxford University Press.

Newsweek. 1984. "Background on Bali." August 23: 84.

Odum, Eugene. 1971. *Fundamentals of Ecology.* Philadelphia: Saunders.

Odum, Howard. T. 1970. *Environment, Power and Society.* New York: Wiley-Interscience.

Oreskes, Naomi, and Erik Conway. 2010. *Merchants of Doubt.* London: Bloomsbury Press.

Osborn, Frederick. 1948–1950. "Forward." In *Studies in Social Psychology in World War II: The American Soldier. Vol. 1, Adjustment During Army Life,* edited by Samuel A. Stouffer et al. New York: Henry Holt.

Oyama, Susan. 2010. "Biologists Behaving Badly: Vitalism and the Language of Language." *History and Philosophy of Life Sciences* 32:401–424.

Palazzoli, Mara Selvini, Luigi Boscolo, Gianfranco Cecchin, and Giuliana Prata. 1978. *Paradox and Counterparadox: A New Model in the Therapy of the Family in Schizophrenic Transaction.* Translated by Elisabeth V. Burt. New York: Jason Aronson.

Pape, Helmut. 1999. "Abduction and the Typology of Human Cognition." *Transactions of the Charles S. Peirce Society* 30.2: 248–269. Accessed April 23, 2004. http://faust15-eth .rz.unifranfurt.de/~wirth/texte/pape.html.

Peirce, Charles S. 1892. "Man's Glassy Essence." *The Monist* 3(1): 1–22.

Perrow, Charles. 1967. "A Framework for the Comparative Analysis of Organizations." *American Sociological Review* 32:194–208.

Pickering, Andrew. 2010. *The Cybernetic Brain.* Chicago: University of Chicago Press.

Price, David. 1998. "Gregory Bateson and the OSS." *Human Organization,* 57(4): 379–384.

———. 2008. *Anthropological Intelligence: The Deployment and Neglect of American Anthropology in the Second World War.* Durham, N.C.: Duke University Press.

Pringle, Robert. 2004. *A Short History of Bali: Indonesia's Hindu Realm.* Crows Nest, New South Wales: Allen and Unwin.

Radcliffe-Brown, A. R. 1957. *A Natural Science of Society.* Glencoe, Ill.: Free Press.

Reich, Charles, 1971. *The Greening of America.* Toronto: Bantam Books.

Reid, Walter V. et al. 2005. Millennial Ecosystem Assessment. Ecosystems and Human Well-being: Synthesis. March 23, 2005. World Resources Institute. Washington, D.C.: Island Press. Accessed May 5, 2005. http://www.millenniumassessment.org/ documents/document.356.aspx.pdf.

Richardson, Lewis F. 1967. *Statistics of Deadly Quarrels.* Ann Arbor, MI: Inter-university Consortium for Political and Social Research.

Rosen, Robert. 1991. *Life Itself: A Comprehensive Inquiry into the Nature, Origin and Fabrication of Life.* New York: Columbia University Press.

Roszak, Theodore. 1969. *The Making of a Counterculture.* Garden City, N.Y.: Doubleday.

Rubinoff, Lionel. 1970. *Collingwood and the Reform of Metaphysics: A Study in the Philosophy of Mind.* Toronto: University of Toronto Press.

Ruiz, Alfredo. 1996. "The Contributions of Humberto Maturana to the Sciences of Complexity and Psychology." *Journal of Constructivist Psychology* 9(4): 283–302.

Ryn, Sim van der. 2013. "Interview." Accessed February 10, 2013. www.vanderryn.com/Work/Projects/Offices/bateson.html

Sahakian, W. H., ed. 1965. *Psychology of Personality: Readings in Theory.* Chicago: Rand McNally.

Sakar, Sahotra. 2006. "From Genes as Determinants to DNA as Resource: Historical Notes on Development and Genetics." In Neumann-Held and Rehmann-Sutter 2006, 77–98.

Salthe, Stanley N. 1985. *Evolving Hierarchical Systems: Their Structure and Representation.* New York: Columbia University Press.

Sapir, Edward. 1994. *Ethnology.* Berlin: Mouton de Gruyter.

Sebeok, Thomas A. 1979. *The Sign and Its Masters.* London: University Press of America.

Secord, Paul, and Carl W. Backman. 1964. *Social Psychology.* New York: McGraw-Hill.

Segal, Lynn. 1986. *The Dream of Reality: Heinz von Foerster's Constructivism.* New York: Norton.

Sluzki, Carlos, and Donald C. Ransom, eds. 1976. *Double Bind: The Communicational Approach to the Family.* New York: Grune and Stratton.

Spies, Walter, and Beryl de Zoete. 2002. *Dance and Drama in Bali.* Hong Kong: Periplus Editions.

Sterelny, Kim, and Paul E. Griffiths. 1999. *Sex and Death: An Introduction to the Philosophy of Biology.* Chicago: University of Chicago Press.

Thibaut, John W., and Harold Kelley. 1971. "Interpersonal Interaction." In Hinton and Reitz 1971, 33–41.

Thompson, Keith Stewart. 1988. *Morphogenesis and Evolution.* Oxford: Oxford University Press.

Thomson, James D. 2010. "Flowering Phenology, Fruiting Success and Deterioration in Pollination in an Early Flowering Geophyte" (September 6). DOI: 10.1098/rstb.2010.0115.

Time Magazine. 1946. "It Is Art: South Sea Spooks." Available from http://www.time.com/time/magazine/article/0,9171,854154,00.html.

Turing, Alan M. 1950. "Computing Machinery and Intelligence." *Mind* LIX(236): 433–460.

———. 1952. "The Chemical Basis of Morphogenesis." *Philosophical Transactions of the Royal Society* 327:37–72.

Turner, Fred. 2006. *From Counterculture to Cyberculture: Stewart Brand, the Whole Earth Network and the Rise of Digital Utopianism.* Chicago: University of Chicago Press.

Uexküll, Jakob von. 2010. "The Theory of Meaning." In *Essential Readings in Biosemiotics,* edited by D. Favareau, 81–114. Dordrecht: Springer.

Vehkavaara, Tommi. 2005. "Limitations on Applying Peircian Semeiotic: Biosemiotics as Applied Objective Ethics and Esthetics Rather Than Semeiotic." *Journal of Biosemiotics* 1(2): 269–308.

Villareal, Luis P. 2012. "The Addiction Module as Social Force." In Witzany 2012, 107–146.

Virilio, Paul. 1995. *The Art of the Motor*. Minneapolis: University of Minnesota Press.

von Neumann, John, and Oskar Morgenstern. 1953. *Theory of Games and Economic Behavior*. 2nd edition. Princeton: Princeton University Press.

Watzlawick, Paul, Janet Beavin, and Don Jackson. 1967. *Pragmatics of Human Communication*. New York. Norton.

Watzlawick, Paul, and John Weakland, eds. 1977. *The Interactional View*. New York: Norton.

Weakland, John. 1967. "Communication and Behavior—An Introduction." *The American Behavioural Scientist* April: 1–3.

———. 1976a. "The Double Bind Theory by Self-Reflexive Hindsight." In Sluzki and Ransom 1976, 307–314.

———. 1976b. "Communication Theory and Clinical Change." In *Family Therapy: Theory and Practice*, edited by Philip J Guerin Jr. New York: Gardner Press, 111–128.

Weaver, Warren. 1966. "The Mathematics of Communication." In *Culture and Communication: Readings in the Codes of Human Interaction*, edited by Alfred G. Smith, 15–24. New York: Holt, Rinehart & Winston.

Webster, Gerry, and Brian Goodwin. 2006. "The Origin of Species: A Structuralist Approach." In Neumann-Held and Rehmann-Sutter 2006, 99–134.

Weizenbaum, Joseph 1976. *Computer Power and Human Reason*. London: W. H. Freeman.

Wilden, Anthony. 1972. *System and Structure: Essays in Communication and Change*. London: Tavistock.

———. 1985. "Context Theory: The New Science" RSSI 5/2 (1986): 97–116.

———. 1987. *The Rules Are No Game: The Strategy of Communication*. New York: Routledge and Kegan Paul.

Wilensky, Harold L. 1967. *Organizational Intelligence: Knowledge and Policy in Government and Industry*. New York: Basic Books.

Wilson, Godfrey, and Monica Hunter. 1945. *The Analysis of Social Change*. Cambridge: Cambridge University Press.

Witzany, Günther, ed. 2012. *Viruses: Essential Agents of Life*. Dordrecht: Springer.

Worster, Donald. 1994. *Nature's Economy: A History of Ecological Ideas*. Cambridge: Cambridge University Press.

Wynne, Lyman. 1976. "On the Anguish and Creative Passion of Not Escaping the Double Bind." In Sluzki and Ransom 1976, 243–250.

Zellmer, A. J., and T. F. H. Allen, and K. Kesseboehmer. 2008. "The Nature of Ecological Complexity: A Protocol for Building the Narrative." *Ecological Complexity* 5:171–182.

Zlatev, Jordan. 2009. "The Semiotic Hierarchy: Life, Consciousness, Signs and Language." *Cognitive Semiotics* 4:169–200.

3. UNPUBLISHED SOURCES

A. The Library of Congress (LOC) in the Margaret Mead Archive (3A[i])

1934. The Eidos of Iatmul Culture. LOC General Correspondence Box 10.

1941. Memorandum on Regularities and Differences in National Character. Council on Human Relations.

c. 1942. Time Perspectives and Topological Psychology: A Comment on Lewin's publication 'Studies in Topological and Vector Psychology.'

1943 (July) 'Commentary' to accompany Bali Exhibit.

1945 (9 January) Weekly Report 1 January–8 January 1945, Arakan Field Unit OSS-SEAC Bittersweet-MO. Gregory Bateson to Carleton F. Scofield.

c.1946–47 Memo. Contributions to Psycho-Cultural Theory. LOC Box 3.

Bateson, G. and Mead, M. Inter Cultural Institute *Bulletin*, LOC 14/1/46 and 13/2/46 Box 03.

Mead, Margaret. LOC 16/07/1940. Macy Exploration, Psychosomatic Medicine Box F 47.

Mead, Margaret. 1941. 'Memorandum on Regularities and Differences in National Character.' Council on Human Relations. Unpublished.

Correspondence, Library of Congress (LOC) Unpublished Letters of Gregory Bateson Consulted or Cited, 1926–47 (3A[ii])

Box	Date	Subject
General Correspondence 01	16 July 1926	to A. C. Haddon
General Correspondence 03	21 April 1947	to Douglas Haring
General Correspondence 01	1 November 1935	Meyer Fortes to Bateson
General Correspondence 01	1/or 3/ November 1935	to Bronislaw Malinowski
General Correspondence 01	5 November 1935	to Bronislaw Malinowski
General Correspondence 02	2 May 1936	to F. C. Bartlett
General Correspondence 02	30 August 1936	to A. C. Haddon
General Correspondence 02	1 October 1936	to F.C. Bartlett
General Correspondence 02	13 October 1936	F. C. Bartlett to Bateson
General Correspondence 01	12 January 1937	from A. R. Radcliffe-Brown
General Correspondence 02	12 January 1937	to F. C. Bartlett
General Correspondence 02	14 February 1937	to) de Navarro
General Correspondence 02	14 March 1937	to A. R. Radcliffe-Brown
General Correspondence 02	29 March 1938	to A. R. Radcliffe-Brown
General Correspondence 01	4 April 1937	to A. R. Radcliffe-Brown
General Correspondence 01	13 April 1938	to W. J. Perry
General Correspondence 02	17 May 1939	to Robert H. Thouless
General Correspondence 02	20 September 1939	to Master, St. John's College
General Correspondence 02	1 October 1940	to (Sir) John Cockcroft
General Correspondence 02	6 November 1940	to Kurt Lewin

Family Correspondence 01	23 September 1940	to Beatrice Bateson
Family Correspondence 01	10 December 1940	to Beatrice Bateson
Family Correspondence 01	7 April 1941	to Beatrice Bateson
Family Correspondence 01	15 March 1942	to Aunt Dick
General Correspondence 02	18 March 1942	to Conrad Waddington
General Correspondence 03	25 June, 1942	J. W. Irving to Bateson
General Correspondence 05	3 October 1943	'Comment on Proposal for Military Police School in Liberated Territories'
General Correspondence 07	18 January 1944	to Clyde Kluckhohn
General Correspondence 03	21 February 1946	Alfred E. Vieres to Bateson
Proposed Conference on Teleological Mechanism	19 March 1946	to Eunice Miner
General Correspondence 03	20 April 1946	to John Henry Hutton
General Correspondence 02	21 August 1946	to A. R. Radcliffe-Brown
General Correspondence 02	11 April 1947	to Douglas Haring

B. The Gregory Bateson archive in the special collections section of the McHenry Library of the University of California, Santa Cruz. This special collection is arranged as follows:

Bk. Mss. (Book Manuscripts ca. 1950–80, boxes 1–9): see 3B(i)

CAF (Complete Articles File): see 3B(ii)

Misc. Mss. (Miscellaneous Manuscripts boxes 1–6): see 3B(iii)

Notebooks and Loose Notes (1943–77 boxes 1–6): see 3B(iv)

Audiotapes (boxes 1–6) and Films: see 3B(v)

MC (McHenry Library Correspondence and Subject Files) (1939–80 boxes 1–39): see 3B(vi)

3B(i) Bk.Mss (Book Manuscripts ca. 1950–80, boxes 1–9)

Bateson, G. 1973. A Way of Seeing. Box 5. Book Manuscripts.

n.d. What Every Schoolboy Knows.

Bateson, G. Abduction. Box 6. (Unlabeled folder marked "Scraps Out").

n.d. Forward [sic] to Ms. "The Evolutionary Idea." McHenry. Book Manuscripts Box 5.

n.d. A Way of Seeing (Ms. Beginning "D; what do you look at?"). McHenry. Book Ms. Box 5. (Bateson, n.d. *Angels Unedited Mss.* no. 2)

Bateson, G. and M. C. Bateson, 1987. *Angels Fear*, Unedited Manuscript Box 5; 1987: 205.27

3B(ii) CAF (Complete Articles File)

1955. Four Lectures. McHenry CAF 126.

1956. Notes on Group Theory. CAF 208.

1971a. The Nature and Culture of Man. McHenry: CAF 201.

1971b. Some Thoughts about Intermittent Reinforcement. McHenry: CAF 308.

1975. Creatural Theory. McHenry: CAF 69/4–5 (dated April 4–6, 1975, no pagination).

Proposal. 1976. Metaphor and Metaphysics. McHenry. ca 1976–77. #955–36a to John Brockman.

3B(iii) Misc. Mss. (Miscellaneous Manuscripts boxes 1–6)

Dialogue between G.B. and Werner Erhard 3 September, 1976 and 6/September, 976, day 2, p. 2.

3B(iv) Notebooks and Loose Notes (1943–77 boxes 1–6)

1949. Notebook 11.

1951. Notebook 14.

1965. Notebook 35.

1967 Notebook 38.

1971 Notebook 45. Route 128: The Jobless Engineers.

1973 Notebook 51/Fall.

1974 Notebook 56/Fall.

3B(v) Audiotapes (boxes 1–6) and Films

1975. Metaphors and Butterflies.

Interfaces—Boundaries Which Connect.

The Pattern Which Connects.

3B(vi) Unpublished Letters of Gregory Bateson Consulted or Cited, 1926–47, McHenry Library, University of California, Santa Cruz (Correspondence and Subject Files) (1939–80 boxes 1–39)

Box	Date	Subject
MC 585–1a	24 April 1947	to Guggenheim Foundation
MC 879–8c	1 February 1949	to Margaret Lowenfeld
MC1039–10b	25 October 1962	to Warren McCulloch
MC858–96a	25 March 1964	to John Lilly
MC 463–4b	11 June 1964	Review of Mss. Evolution
MC 1399–1c	18 May 1965	to V.C. Wynne-Edwards
MC 1024	15 Nov 1965	'Foreward' [sic] (unpublished)
MC 876–1	2 May 1966	to Konrad Lorenz
MC 876–3	7 August 1966	to Konrad Lorenz
MC 1068–2a	12 December 1966	to William Parker

MC 1068–5	19 February 1967	from William Parker
MC 1519-1b	1 June 1967	to Philip Wylie
MC 672–17	22 February 1968	to Tolly Holt
MC 858–144	10 May 1968	to John Lilly
MC 1056	28 July 1968	to Lita Osmundsen
MC 1019–63a	23 January 1969	to Ken Norris
MC 488–3	ca. May 1969	Comments on Film Proposal
MC 52–6	23 June 1969	to Robert Ryder
MC 1497–7a	21 October 1969	from Anthony Wilden
MC 733–2	11 November 1969	to Will Jones
MC 1519–8e	12 May 1970	to Philip Wylie
MC 593–103b	7 September 1970	Jay Haley Mss. "Development of a Theory"
MC 771–2a	11 September 1970	to John Y. Kim
MC1209–11	4 March 1972	to Rev. Riyadassa
MC 824–2	15 March 1973	to Edmund Leach
MC 1400–9b	30 May 1974	to Donald Campbell
MC250-7b	11November 1974	to Mary Scott Dewire
MC 955–36	ca.1976	to John Brockman
MC 559.5–2a	11 June 1979	to John Goppelt
MC127–5b	4 January 1980	to (Sir) Patrick Bateson

4. VIDEOS

Ben-Jacob, Eshel. *Social Intelligence of Bacteria: Learning from Bacteria About Social Networks*. Retrieved September 30, 2011, from https://www.youtube.com/watch?v=yJpi8SnFXHs.

Hansen, James, Natalia Shakova, Peter Wadhams, and David Wasdell. 2013. *Arctic Methane: Why the Sea Ice Matters*. Retrieved February 9, 2013, from www.youtube.com/watch?v=iSsPHytEnJM You Tube.

McNeil, Don. 2010. "Going on with Systems." Unreleased.

The Silence of the Bees. 2007 (October). Retrieved from www.pbs.org/wnet/nature/episodes/silence-of-the-bees/video.

Suzuki, David. 2012. *Smarty Plants: Uncovering the Secret World of Plant Behaviour*. Retrieved September 27, 2012, from www.cbc.ca/natureofthings/episode/smarty-plants-uncovering-the-secret-world-of-plant-behaviour.html.

The Vanishing of the Bees. 2009. Retrieved from www.imdb.com/title.

INDEX

behaviorism, 4, 28, 62, 75, 89, 135, 253n1

Belo, Jane, 36, 113

Benedict, Ruth, 10, 49–51, 54, 64

binary (or collateral) comparisons: analog/digital, 137, 139, 190, 200–204, 246–247; annular/meridial, 189–190; body/mind, 10, 13, 32–35, 90, 207; Creatura/Pleroma, 161–164; ethos/eidos, 25–26, 29, 32–33, 35–36, 47, 66, 68, 208, 224; fragmented (single) vision/*moiré*, xi, 13, 68, 228; information/matter, 105, 123, 150, 161, 244–248; inside/outside, 13, 105, 140, 159–161, 170, 192, 206, 239; linearity/nonlinearity, 1, 3, 14, 70, 86–87, 90, 95, 118, 137, 188, 199, 229, 244–246, 248; monism/dualism, 10; objectivity/subjectivity, 11, 116–117, 185–186, 210, 212; synchronic/diachronic patterns, 10, 51, 67. *See also* nonlinearity

bioentropy, 150–155, 157–158, 257n2

biology, 3, 89, 128, 148, 150–151, 180, 183, 204, 207; and approach to consciousness, 221; Bateson's background in, 6; bioaccumulation, 145–146; biodiversity, 149, 169, 234–235, 241; molecular, 2, 6, 12, 139, 163, 170, 175–176, 202; and natural selection, Darwinism, 155–156. *See also* developmental biology

biosemantics, 12, 193; and meaning rationalism, 195–201

biosemiotics, 12, 141, 185, 193–194, 206–207, 212–216, 241; code duality in, 139, 201–205; endosemiotics, 202; semethic interaction, 208

biosphere, 2, 12, 151, 180, 194; reserves, 235

Blake, William, 13, 101, 127, 223, 237. *See also* binary (or collateral) comparisions: inside/outside

Bolter, J. D., 137, 139

Boulding, Kenneth, 81

boundaries, 2, 13, 28, 87–88, 103, 123–125, 164, 174–175, 178, 185, 192, 212, 222; functional, 205; and moving boundary problem, 179; and schizophrenia, 114–115; and sign use, 196, 198

Bourdieu, Pierre, 25

Brand, Stewart, 93, 95–96

Bredo, Eric, 85, 109

Brown, Governor Edmund G. (Jerry), 94, 231

butterflies, 17, 241; bread and butterfly fable, 168–169; population depletion, 154, 237–239

butyric acid, 157, 184. *See also* von Uexküll, Jacob

California, 238–239; Esalen Center, 95–96; Gregory Bateson Building, Sacramento, 231; Santa Cruz (UCSC), x, 9, 66, 93–94, 146; Veterans Administration Hospital, Palo Alto, 9, 76; War Resisters League, 95; Zen Center, San Francisco, x

Campbell, Jeremy, 72–73

Canada, 48, 236–237, 258n1

Carroll, Lewis, 168

Carson, Rachel, 145, 147

citizen scientists, 241

climate change, 133, 231–234, 239, 258nn1–2; Copenhagen Conference on Climate Change (2009), 234; global warming, 5, 16, 231–234, 239–240; International Panel on Climate Change (IPCC), 16, 232–234

Cock, Alan, 6, 253n1

coding transforms, 140, 210, 212, 222

coevolution, 139, 155–156, 166–167, 192, 217, 228

coevolution and morphogenesis, 169–174

cognitive semiotics, 213–214

Collingwood, R. G., 13, 15, 225–227

communication, 53, 84–86, 118, 213–214; animal, 12, 199, 202; and digital coding, 204; and double bind, 119–122; in families, 111–113, 125–126, 137; and feedback and information theory, 2, 8–9, 67, 75; in living systems, 136, 157–158, 170, 194; and logical typing, 116–119; and metalinguistic ability, 257n1; octopuses, 201; and patterns in interpersonal communication, 103–105; and schizophrenia, 113–116, 130; and Shannon-Weaver concept of information, 73, 243–244; and sign usage, 197; as transaction, 81–84

MEANING SYSTEMS